REPORT

An Assessment of the Assignments and Arrangements of the Executive Agent for DoD Biometrics and Status Report on the DoD Biometrics Enterprise

Douglas Shontz • Martin C. Libicki • Rena Rudavsky • Melissa A. Bradley

Prepared for the Office of the Secretary of Defense

Approved for public release; distribution unlimited

RAND NATIONAL DEFENSE RESEARCH INSTITUTE

The research described in this report was prepared for the Office of the Secretary of Defense (OSD). The research was conducted within the RAND National Defense Research Institute, a federally funded research and development center sponsored by OSD, the Joint Staff, the Unified Combatant Commands, the Navy, the Marine Corps, the defense agencies, and the defense Intelligence Community under Contract W74V8H-06-C-0002.

Library of Congress Control Number: 2012951038

ISBN: 978-0-8330-7707-3

The RAND Corporation is a nonprofit institution that helps improve policy and decisionmaking through research and analysis. RAND's publications do not necessarily reflect the opinions of its research clients and sponsors.

RAND® is a registered trademark.

© Copyright 2012 RAND Corporation

Permission is given to duplicate this document for personal use only, as long as it is unaltered and complete. Copies may not be duplicated for commercial purposes. Unauthorized posting of RAND documents to a non-RAND website is prohibited. RAND documents are protected under copyright law. For information on reprint and linking permissions, please visit the RAND permissions page (http://www.rand.org/publications/permissions.html).

Published 2012 by the RAND Corporation
1776 Main Street, P.O. Box 2138, Santa Monica, CA 90407-2138
1200 South Hayes Street, Arlington, VA 22202-5050
4570 Fifth Avenue, Suite 600, Pittsburgh, PA 15213-2665
RAND URL: http://www.rand.org/
To order RAND documents or to obtain additional information, contact
Distribution Services: Telephone: (310) 451-7002;
Fax: (310) 451-6915; Email: order@rand.org

Preface

Biometrics—the identification of humans based on the characterization of physical traits—have assumed an increasing role in the Department of Defense's (DoD) operations since 2001. Similar to the U.S. experience in Vietnam, the wars in Afghanistan and Iraq presented particular challenges because combatants did not wear uniforms and blended into the local population. However, advances in biometric technology provided potential ways to increase the certainty about a person's identity and connections to other people and events.

DoD biometrics activities have thus far been funded overwhelmingly by overseas contingency operations (OCO) funds, and technology has been procured quickly in response to urgent needs from the field. However, with U.S. operations ended in Iraq and major operations ending in Afghanistan, DoD must decide what to preserve of its biometrics enterprise, how to manage it, and what the future enterprise will be expected to do.

The Principal Staff Assistant (PSA) for Biometrics within the Office of the Secretary of Defense (OSD) asked RAND to study the biometrics enterprise. Specifically, the PSA asked RAND to assess the assignments and arrangements of the Executive Agent (EA) for biometrics—currently the Secretary of the Army (SECARMY)—provide a status report on the entire biometrics program for the Secretary of Defense, and provide the PSA options and recommendations to help future planning.

This research was sponsored by the PSA and conducted within the Acquisition and Technology Policy Center of the RAND National Defense Research Institute, a federally funded research and development center sponsored by the OSD, the Joint Staff, the Unified Combatant Commands (COCOMs), the Navy, the Marine Corps, the defense agencies, and the defense Intelligence Community.

For more information on the RAND Acquisition and Technology Policy Center, see http://www.rand.org/nsrd/ndri/centers/atp.html or contact the director (contact information is provided on the web page).

Contents

Figures and Tables

Figures

Tables

Summary

Background

In historical combat and in accordance with the Laws of Armed Conflict, soldiers were easily identified because they wore uniforms or bore other easily distinguished markings. However, the United States faced foes in Vietnam, after the terrorist attacks of September 11, 2001, and in two subsequent wars in Afghanistan and Iraq, who attempted to blend in with the population as much as possible. The need to identify known and suspected combatants and bad actors has elevated the importance of biometrics in the Department of Defense (DoD). Specifically, biometrics provide a potential way to increase the certainty about a person's identity, record of previous encounters, and connections to other people and events. The field of biometrics is founded on the observation that certain physical characteristics can differentiate one person from another and do so in a way that is stable over time.

In DoD, biometrics have been increasingly used to identify enemies (red force) and to verify friendly (blue force) and other (gray force) personnel. The Secretary of the Army (SECARMY) is the Executive Agent (EA) of the DoD Biometrics Program, which has largely been supported by overseas contingency operation (OCO) funds to date. However, the withdrawal of U.S. forces from Iraq and the pending withdrawal of combat forces from Afghanistan by 2014 means OCO funds are being cut. Additionally, questions have been raised about the efficiency and effectiveness of current biometrics activities.

Purpose

The Principal Staff Assistant (PSA) asked RAND to:

1. Assess the assignments and arrangements the EA put in place to meet user needs;
2. Provide a status report on the biometrics enterprise to the Secretary of Defense; and
3. Provide the PSA with options and recommendations for the biometrics enterprise to help develop input for the Fiscal Year 2014 (FY14) Program Objective Memorandum (POM).

Study Methodology

The report is based on research conducted from September 2011 through January 2012 that includes information on current activities and historical events. Because the biometrics program lacks quantifiable objectives, RAND researchers relied on qualitative data derived from official documents and interviews with DoD and other personnel.

Official documents show the EA's original purpose of creating a standard business process to be "pushed out" to the DoD components shifted to the current biometrics program in which the EA is expected to "pull" user requirements and respond with tactical devices supported by a common architecture. However, guiding documents have not clearly defined the end-users, nor have these documents established a process for prioritizing or standardizing requirements for multiple DoD components. Complicating any attempt to conduct an assessment is the lack of agreement about whether the biometrics program is responsible to "push" capability based on the program's own volition, or "pull" requirements from users, and to what end.

As a result, to conduct the assessment in terms of end-user needs, RAND researchers examined the EA's assignments and arrangements from the perspective of the biometrics cycle, i.e., the steps used to collect and use biometric data. This provided RAND a way to understand the EA's role and identify potential risks to a functioning biometrics capability.

To complete the status report, RAND researchers looked at the full scope of DoD biometrics work and identified gaps in policy, implementation, and other areas that potentially pose risks to a functioning enterprise.

Assessment of the Assignments and Arrangements of the EA

The EA is supporting a functioning biometrics cycle that meets some level of warfighter needs and has had significant success identifying bad actors. Aspects of the EA's assignments and arrangements show some deficiencies in managing biometrics responsibilities and pose risks to the biometrics cycle and ability to provide for user needs. Further, this assessment examines ten years of activity and was conducted at a single point in time. Thus, it does not fully capture the recent efforts by the EA to improve operations, many of which were ongoing during our research. The EA has worked to increase automation, formalize activities, and develop metrics that demonstrate the impact of biometrics work. RAND expects that the required future annual assessments will track these continued changes and efforts to improve effectiveness and efficiency.

EA Management

With respect to how well the EA and the EA's components are carrying out their functions, several aspects of that management could be reducing the effectiveness and efficiency of biometrics for users. Management deficiencies appear to stem from three causes. First, EA responsibilities are "stovepiped" within the Army, and the Executive Manager is not empowered to exert management authority. The EA's responsibilities are spread across the Army and meet only at the level of SECARMY, who is designated the EA by DoD Directive 8521.01E. Most EA functions are delegated to Army components, and it is infeasible for the EA himself to manage the program.

Second, most of the biometrics activities have been supported by OCO funding and have focused on procuring quick reaction capabilities (QRCs) because of the demands of the wartime environment. The absence of structure and strategic planning contributed to the proliferation of devices that are not completely interoperable and to conflicts between end-users and EA components over which tasks were given priority.

Third, the EA lacks metrics to determine the sufficiency and efficiency of its activities. DoD as a whole lacks goals for using biometrics and implementing biometric technology. This contributes to deficiencies, such as a Biometrics Enterprise Strategic Plan (BESP) that contains general goals and objectives that cannot be measured. Nevertheless, the EA is still responsible for developing organizational goals and objectives to provide guidance and enforce accountability.

EA Support of the Biometrics Cycle and End-User Needs

The EA is working to ensure biometrics are being collected and have been used to identify individuals against prior encounters and forensic evidence. DoD's authoritative database is growing steadily, and the processes in the field are becoming increasingly routinized. However, an examination of the EA's assignments and arrangements suggests risks and room for improvement.

The biometrics cycle can be characterized as the following series of steps that explain the movement and use of biometric data:

> Task/Direct —> Collect —> Transmit —> Match —> Store —> Reference —> Analyze —> Decide/Act (—> back to Task/Direct).

In the absence of concrete goals for the EA's support to end-users, RAND researchers assessed the EA's effect on the cycle. Specifically, RAND researchers considered ways the cycle could fail (referred to as "failure modes") and how EA assignments and arrangements could pose risks to the cycle now and in the future. This report discusses issues that RAND researchers assessed as presenting the greatest risks to the functioning of the biometrics cycle. The intent is to highlight areas where the EA could make changes that might improve how the cycle functions, in turn improving effectiveness and efficiency for end-users.

Status of the Biometrics Enterprise

The greatest challenge in making statements about the status of the biometrics enterprise is that it lacks clear goals and measurable objectives against which we could judge its health and performance. Consequently, RAND researchers considered gaps in policy, implementation, and related areas, and drew conclusions about the overall status across DoD.

In general, DoD's biometrics capability regarding red and gray forces exists, and has been supporting warfighter needs to some extent. The most frequent complaints in personnel interviews were about inefficiencies in moving biometric data, lack of feedback to the field, lack of capabilities spanning the full range of missions and environments, and EA management issues. Changes to the current scheme are in the works but will not be fielded before FY15 at the earliest.

A major reason for the current state of biometrics capabilities is that there is no common understanding across DoD about the purpose and boundaries of the enterprise. DoD also lacks specific and measurable strategic goals and objectives for using biometrics. In other words, DoD has no way to evaluate the sufficiency of the biometrics enterprise. A clearly defined set of goals, objectives, and boundaries for the enterprise would enable better discussion of the issues raised throughout this study.

The structural inefficiencies and lack of goals have had a direct effect on the current status. DoD's biometrics capabilities have not transitioned from QRCs to a coherent set of programs and activities, despite apparent support for long-term use. Many of the disputes about maintaining and transitioning the QRCs will be resolved de facto by the decline of OCO funding and expected DoD-wide budget cuts.

DoD has made some effort in blue-force biometrics activities, but this is still in very preliminary stages, with many unresolved questions. Most important, shrinking budgets will require a clear justification for any new investments, and this justification must include cost-benefit analyses, risk assessments, and discussions about trade-offs.

This report discusses gaps that RAND researchers assessed as presenting the greatest risks to the biometrics enterprise. Similar to the assessment of the EA's assignments and arrangements, the intent is to highlight areas where DoD could make changes to potentially improve the enterprise. Also similar to the assessment of the EA, the biometrics enterprise has worked in recent years to formalize activities and make biometrics more of an institutional capability. RAND expects that the required future status reports will track these continued changes and efforts to improve the enterprise's effectiveness.

Conclusions and Recommendations

Conclusions

DoD's biometrics systems function as a result of the hard work of many people, and these systems have helped identify combatants and bad actors. However, after ten years, DoD lacks an agreed-upon set of goals for biometrics use and a coherent end-to-end capability. The EA shifted from "pushing" technology to DoD components to "pulling" requirements and responding with tactical biometric equipment. Further, the biometrics enterprise lacked strategic focus, which prevented it from laying a path to a coherent, integrated, and cost-effective system.

The way forward is unclear. DoD's biometrics activities grew and were executed from the bottom up primarily to support a large conventional force engaged in mass collection of people's information in wartime. However, all indications right now are that biometrics' use in the future will be focused on intelligence, counterterrorism, other threat networks, interagency law enforcement support, and international partner cooperation. This means the biometrics enterprise will have to make the case that biometrics provide the most cost-effective approach for addressing designated missions and threats, based on a firm analytical foundation beyond anecdotes of success from the field. DoD must decide both what is truly needed for and expected of an end-to-end biometrics capability. Key to this will be DoD and the EA developing outcome metrics for assessing the value, effectiveness, and efficiency of biometrics activities. Metrics are regularly cited as critical for DoD's wide range of activities, and the field of biometrics is no different.

Options and Recommendations

Based on what RAND researchers see as unresolved issues in the enterprise and on apparent future needs, this report includes a set of options and a set of recommendations related to the EA and DoD writ large.

Options: RAND researchers presented a set of options for DoD to consider with respect to the EA, information-sharing, and development and operation of the authoritative biometrics database of red and gray forces.

First, when examining the EA, DoD should consider whether that role should be modified and whether a different DoD component may be better suited to providing for users' future biometrics needs, given the expected reduction in conventional, ground-force collection of biometric data after Afghanistan operations end.

Second, when considering information–sharing and interoperability, DoD should look for ways to increase the amount of data shared with interagency partners, including the Department of Homeland Security (DHS) and the Intelligence Community (IC).

Third, when considering efficient operation and use of biometric data, DoD should examine the current division between developing the authoritative database and conducting day-to-day biometrics data operations.

Recommendations: RAND researchers also presented a set of specific recommendations that they believe could improve near-term biometrics enterprise management and activities.

First, RAND recommends DoD issue a clear and understandable articulation of the mission for the biometrics enterprise that, at a minimum, replaces the current BESP. This should cover red-, gray-, and blue-force biometrics; contain an explicit statement about biometrics being an enabling and support tool for warfighter missions; emphasize the need for a functioning end-to-end biometrics capability; and require compliance with biometric data standards. Stating goals for the use of biometrics will help establish boundaries for most future discussions and, in many ways, eliminate long-standing conflicts within DoD.

Second, RAND recommends Department of Defense Directive (DoDD) 8521.01E be revised to clarify the objective of the biometrics EA with respect to a complete and functioning end-to-end biometrics cycle. Within this framework, RAND recommends the updated DoDD ensure that roles and responsibilities have been delineated to confirm responsibility for the cycle and its components and to clarify mediation processes for organizational conflicts. Among other things, the revised DoDD 8521.01E should require determination of what constitutes biometrics cycle sufficiency and the reasons for breakdowns in the biometrics cycle when they occur.

Third, RAND recommends DoD eliminate the parallel architectures currently used and consider a standard architecture and scheme that can be used to transmit biometric data and to match responses. This should include: (A) prioritizing implementation of automated classified-to-unclassified data transfers, and (B) ensuring that match/no match responses are sent to the originator of the request as well as other users of the information.

Finally, we recommend that the PSA and EA develop metrics to assess both the outcomes and outputs of the biometrics enterprise. Metrics are an oft-discussed issue in DoD, and the biometrics enterprise needs to improve its use of them. DoD should consider a cost-benefit approach to outcome metrics for biometrics that includes comparing current outlays to current positive results. DoD should also consider more user-focused measures of speed, reliability, and quality of biometric data as compared to some benchmarks for output metrics. The PSA and EA should work together to use and improve output metrics already being used by the EA.

The DoD biometrics enterprise is at a point of significant change, with large-scale OCOs ended (in Iraq) or ending (in Afghanistan) and funding reductions approaching. At the same time, intelligence, law enforcement, and international partner desires for biometric capabilities are likely to increase. DoD must now establish goals for the biometrics enterprise, justify the scope and scale of biometrics-related activities to senior leadership, and ensure the enterprise meets user needs effectively and efficiently.

Acknowledgments

The authors wish to thank the study sponsor and all DoD personnel who agreed to be interviewed, provided information, and reviewed and commented on the draft version of this report to ensure an accurate product. The authors also wish to thank Victoria Greenfield, Peter Higgins, and Katharine Webb for their helpful reviews and comments.

Abbreviations

ABIS	Automated Biometric Identification System
ADM	Acquisition Decision Memoranda
AFRICOM	U.S. Africa Command
ACAT	Acquisition Category
AOR	Area of Responsibility
AROC	Army Requirements Oversight Council
ASA(ALT)	Assistant Secretary of the Army for Acquisition, Logistics and Technology
BAT	Biometrics Automated Toolkit
BEC	Biometrics Enabling Capability
BEI	Biometrically Enabled Intelligence
BESP	Biometrics Enterprise Strategic Plan
BEWL	Biometrically-Enabled Watchlist
BFC	Biometrics Fusion Center
BIMA-CC	Biometrics and Identity Management Agency–Crystal City
BIMA-WV	Biometrics and Identity Management Agency–West Virginia
BISA	Biometric Identification System for Access
BMO	Biometrics Management Office
BSWG	Biometrics Standards Working Group
BTF	Biometrics Task Force

CAC	Common Access Card
CAPE	Cost Assessment and Program Evaluation
CDD	Capability Development Document
CENTCOM	U.S. Central Command
CJIS	Criminal Justice Information Services (DOJ-FBI database owner)
COCOM	Combatant Command
CONOP	Concept of Operations
COOP	Continuity of Operations
CPD	Capability Production Documents
DCR	Doctrine, Organization, Training, Materiel, Leadership and Education, Personnel, and Facilities Change Recommendation
DHS	Department of Homeland Security
DIA	Defense Intelligence Agency
DISR	DoD Information Technology Standards and Profile Registry
DMDC	Defense Manpower Data Center
DoD	Department of Defense
DoDD	DoD Directive
DoDI	DoD Instruction
DOJ	Department of Justice
DOS	Department of State
EA	Executive Agent
EBTS	Electronic Biometric Transmission Specification
EUCOM	U.S. European Command
FBI	Federal Bureau of Investigation
G-2	Army Deputy Chief of Staff for Intelligence
G-3/5/7	Army Deputy Chief of Staff for Operations, Plans, and Training
G-8	Army Deputy Chief of Staff for Resource Management
GAO	Government Accountability Office

HIIDE	Handheld Interagency Identity Detection Equipment
IA	Information Assurance
IAFIS	Integrated Automated Fingerprint Identification System
IC	Intelligence Community
IDA	Institute for Defense Analyses
IDENT	DHS Automated Biometric Identification System
IdM	Identity Management
IdMSCG	Identity Management Senior Coordination Group
IDS	Identity Dominance System
IED	Improvised Explosive Device
INSCOM	Intelligence and Security Command
J8	Joint Staff Force Structure, Resources, and Assessment Directorate
J34	Joint Staff Deputy Directorate for Antiterrorism/Homeland Defense
JROC	Joint Requirements Oversight Council
JPIv	Joint Personnel Identification
JUONS	Joint Urgent Operational Needs Statements
LTM	Last Tactical Mile
MIP	Military Intelligence Program
NGIC	National Ground Intelligence Center
NIP	National Intelligence Program
NIPRNet	Non-Classified Internet Protocol Router Network
NIST	National Institute of Standards and Technology
NORTHCOM	U.S. Northern Command
NSS	National Security Staff
O&M	Operations & Maintenance
OCIO	Office of the Chief Information Officer
OCO	Overseas Contingency Operations

OPMG	Office of the Provost Marshal General
OSD	Office of the Secretary of Defense
OSD-CIO	Office of the Secretary of Defense–Chief Information Officer
OUSD(I)	Office of the Under Secretary of Defense for Intelligence
OUSD(P)	Office of the Under Secretary of Defense for Policy
OUSD(P&R)	Office of the Under Secretary of Defense for Personnel and Readiness
PACOM	U.S. Pacific Command
PEO-EIS	Program Executive Office for Enterprise Information Systems
PIN	Personal Identification Number
PM-Biometrics	Program Manager for Biometrics
POM	Program Objective Memorandum
PoR	Program of Record
PSA	Principal Staff Assistant
QRC	Quick Reaction Capability
ROI	Return on Investment
SECARMY	Secretary of the Army
SEEK II	Secure Electronic Enrollment Kit version II
SIGINT	Signals Intelligence
SIPRNet	Secret Internet Protocol Router Network
SOCOM	U.S. Special Operations Command
TCM-B&F	TRADOC Capability Manager for Biometrics and Forensics
TRADOC	U.S. Training and Doctrine Command
US-VISIT	U.S. Visitor and Immigrant Status Indicator Technology
USFOR-A	U.S. Forces Afghanistan

CHAPTER ONE

Introduction

Background

Biometrics have assumed an increasing role in the Department of Defense's (DoD) operations over the last ten years after being initially deployed in Iraq to help manage detainees. Part of the reason for this increased role lies in the changed nature of today's wars. In historical combat and in accordance with the Laws of Armed Conflict, soldiers were easily identified because they wore uniforms or bore other easily distinguished markings.[1] Their exact identities were secondary: The uniform said everything about how they could be treated on the battlefield and thereafter. However, the United States faced foes in Vietnam, after the terrorist attacks of September 11, 2001, and during two subsequent wars in Afghanistan and Iraq who did not wear uniforms; indeed, their *modus operandi* was to blend in with the population as much as possible. This tactic poses difficulties for U.S. forces to determine who they are encountering, e.g., has a person been encountered previously and characterized as potentially hostile? Equally important was the need to know whether a person was associated (in time or location) with other hostile events or "bad actors." Biometrics provided a way to achieve a previously unavailable degree of certainty about a person's identity, record of previous encounters, and connections to other people and events.

In addition to the military's need to accurately identify a large number of people, biometric technology has also improved enough to enable greater use. Rapid advances in acquiring, processing, transmitting, and storing information now allow military forces to ascertain, *within minutes*, whether someone they encounter has been encountered before—and do so for tens of thousands of people a day.

---

1 See, e.g., Article 4, Convention (III) Relative to the Treatment of Prisoners of War, Geneva, August 12, 1949.

Role of Biometrics

Use of biometrics is founded on the observation that certain characteristics of individuals differentiate one person from another in a way that is stable over time. Collecting these biometrics at two points in time can, ideally, establish whether a person encountered today is the same person encountered at some point in the past—and is no one else. The more popular biometrics include fingerprints, facial images, irises, and DNA, although there are many other biometrics in use or mooted for use.[2] The desirable qualities of a biometric include distinguishability, stability over time and circumstance, ease of collection, ease of processing (both by humans and computers), universal acquisition (everyone can generate one), deep archives, and forensic capabilities. Each biometric modality has limits in terms of one or more of these ideal qualities.

Fingerprints are one of the oldest biometric modalities in widespread use, with the most robust technical and analytical infrastructure for their use. No two individuals have the same fingerprints, and fingerprints remain the same over an individual's life, barring scarring or disease, which allows for matching fingerprints taken at different times. Fingerprints have the added feature of being frequently left behind at crime scenes (in the form of a latent fingerprint), allowing investigators to determine with a greater or lesser degree of reliability that a specific individual touched an item or visited a location at some point in the past.

Fingerprints have high distinguishability and stability, useful forensic properties, are easy to computer-process, and are deeply archived, but they must be collected through direct contact by trained personnel to ensure readable prints. A small percentage of the population lacks well-defined fingerprints as well. Irises have better distinguishability and believed stability, and are easy to process, but not that easy to collect; archives of iris images are still small, and no one leaves them behind at the scene of a crime. Facial images have the advantages of easy (even surreptitious) collection, the most widespread archives (more so than fingerprints), and universal possession. They can also be processed by humans to some extent, but the technology to distinguish one face from another (when the sample is very large) is still evolving and, again, they are never left behind, except for images captured by surveillance cameras with varying degrees of clarity. DNA is distinguishable, highly stable, universal, is often left behind at the scene of the crime (sometimes when fingerprints are not), and can be easily processed by computer but not at all by humans. However, collection requires proper training because DNA degrades at different rates depending on environmental conditions and can be easily contaminated.

Because no biometric is perfect, collecting multiple biometric modalities, sometimes in concert with biographic information, can increase the probability of a match and reduce the risk of an incorrect match. Multimodal collection when an individual is fully "enrolled" in a biometric database can also subsequently provide the means for different verification measures. For example, if a person is added to a biometric database using only an iris scan, there is no

[2] Special mention should be made of signatures and speaker recognition (previously referred to as "voiceprints"). Neither provides particular differentiation power nor stability over time and circumstance, but each has a unique role. The signature, as a biometric, can only be collected voluntarily and thus connotes assent to collection. Speaker recognition may be the only biometric that can be collected in some circumstances.

possibility of later matching that person's fingerprints discovered in another location. DoD's handheld biometrics collection device, for example, is made to collect fingerprints, irises, and facial images (but not DNA).

Uses of biometrics generally can be divided into two categories: verification and identification. Biometrics can be used to verify the identity of friendly forces, known as blue forces, and third-party foreign nationals, known as gray forces, for purposes of facility access and other actions.[3] To use an oft-repeated example, a person asserts that he has a right to enter federal property, and passes an identification card forward that has his picture (a biometric). The guard examines the credentials to link authentication with name (establishing that someone with that name and face is in the database), and the picture so that it may be compared with the face of the person holding the credential. In more secure locations, a person may present a fingerprint (or a palm print) so as to validate, with higher degrees of accuracy, that the presenter is who he or she claims to be. The DoD's Common Access Card (CAC) has an embedded biometric, a fingerprint, but facilities and computer access devices that read this biometric as a condition of access are the exception, not the rule.

Biometrics are also used to identify enemy forces, known as red forces, by matching information to previously collected data or through forensic work.

A canonical vignette illustrating the former would be when U.S. forces detain an individual in Afghanistan, collect his biometrics, and learn he is on the watch list, known as the Biometrically-Enabled Watchlist (BEWL). A common example from the wartime environment of forensic identification is retrieving a latent fingerprint from an improvised explosive device (IED) and matching it to an individual who is detained for other reasons. It is important to note that these categories of uses are general. For example, the identities of red forces may be verified for purposes of detainee management and movement. Further, gray forces may go through an identification process to ensure a person has not been previously characterized as hostile. As will be explained in greater detail below, blue-force biometrics remains a limited line of work for DoD, so this report focuses on red- and gray-force biometrics.

Biometrics are only potential enablers in the process of characterizing and validating an individual's identity, i.e., ascertaining that a person is who he says he is (blue biometrics) or accurately identifying someone as hostile (red biometrics). This broader context is sometimes referred to as identity operations, or identity management. Identity management, at some level, is carried out by all organizations as a necessary part of determining who merits what privileges. The process of authenticating and identifying people, depending on context, can take many forms—humans, after all, have been doing it since well before the first fingerprint was collected. For instance, computer network authentication may include a token (something you have, such as a CAC), a password (something you know), and biometrics (something you are). In other circumstances, people can be authenticated by networks of trust, e.g., personal

---

[3] For purposes of this report, blue forces are considered U.S. personnel and allied personnel authorized to enter U.S. or international coalition facilities. Gray forces are local nationals or other foreign nationals who have not been characterized as friendly or enemy, or who are allowed access to certain areas of U.S. or international facilities, but with fewer privileges.

knowledge of an individual's identity based on prior contact, biographical detail, and patterns of behavior. Although our study may contribute to an understanding of identity operations and related issues, we are not examining identity operations writ large.

Modern biometric technology, including that used by DoD, has allowed the process of collecting, storing, and analyzing data to be faster and increasingly automated. The collection device creates a digital image and/or representation of the physical characteristic. The biometric data are transmitted to a database where computer algorithms seek a match to existing records. Depending on the quality of the data collected and the size of the database, this automated attempt may result in several possible matches that are analyzed manually to make a final determination. The newly collected data are stored in the database and the match/no match result is sent to a person who can use the information. Ideally this happens quickly and includes the person who collected the data (who may still be with the person whose data were collected).

An example of this process begins with a person who has been arrested. His fingerprints are collected and transmitted to the Federal Bureau of Investigation's (FBI) national database, where they are successfully matched with fingerprints in a prior criminal record (perhaps from another state). The result is communicated back to the law enforcement agency, which can then decide whether to take additional action against the individual arrested.

The various components (relating both to people and technology) involved in this process ultimately determine the quality of the data collected, the speed of data transmission, whether the data can be analyzed automatically and compared to other data, and the accuracy of the analysis.

DoD Biometrics Program Documentation and the Executive Agent

Since a 2000 DoD memorandum designated the Secretary of the Army (SECARMY) as the Executive Agent (EA) for DoD Biometrics, DoD has sought coordinated deployment of biometric technology among its components.[4] The initiation of a formal DoD Biometrics program occurred as advancements in computing technology supported aspirations of highly integrated identification capability. However, beyond a goal of mass technology deployment, the aim of biometrics capability remained effectively undefined. A series of memos from 2003 to 2005 reveal early variability in goals regarding the biometrics program. In 2003, DoD designated the DoD Biometrics Program an Acquisition Category (ACAT) 1AM in an attempt to elevate the role of biometrics.[5] Shortly after that, in 2004, DoD repealed this designation noting a lack of progress, and conceding a need for a more coherent framework in which to

---

[4] Rudy de Leon, Deputy Secretary of Defense, "Executive Agent for the Department of Defense (DoD) Biometrics Project," memorandum, Washington, D.C., December 27, 2000.

[5] John Stenbit, Pentagon Chief Information Officer, "Designation of Biometrics as ACAT 1AM," memorandum, Washington, D.C., May 15, 2003.

deploy biometric technology; this memo suggests such capability should support an identity management vision, and describes the creation of a group that would develop it.[6] At the same time, troops deployed to Afghanistan and Iraq, as well as counterterrorism units, requested significantly different technological capabilities to complete their missions. Recognizing these urgent operational needs, DoD released a policy requiring biometric screening of individuals in Iraq for base access in 2005.[7] This static biometric collection and match requirement corresponded with another to collect biometric identifiers of the rapidly growing population of Iraqi detainees held by U.S. forces. These requirements and others like them clarified mission needs statements for biometric technology, albeit in very different forms than initially envisioned for mass identity management. In Iraq, biometric collection and matching efforts required electronic messages to traverse networks of different classification levels (Non-Classified Internet Protocol Router Network [NIPRNet] to Secret Internet Protocol Router Network [SIPRNet]). In Afghanistan—a very austere environment where troops regularly travel far from any Internet access—biometric collection devices must be "tactically" designed and oriented to take into account free movement of known terrorists or enemy combatants.

As DoD deployed biometric technology to support urgent warfighter requirements, the department worked to release documentation normally preceding development of such a largely utilized technology. In February 2008, DoD updated the Biometrics EA Directive to establish new roles and responsibilities. This document was followed shortly thereafter by the DoD release of the "Biometrics Enterprise Strategic Plan: 2008–2015," in August 2008, which recognizes the "dynamic" development of the newly named Biometric Enterprise.[8] It also focused the enterprise on the acquisition of technology geared toward "Red Biometrics" and efforts to thwart non-traditional adversaries.[9] The strategic plan's goals are provided in Appendix A.

The chronology of document development for the biometrics enterprise continued asynchronously. In April 2008, prior to releasing any identified, coordinated, and signed joint DoD technology requirements, DoD designated the DoD Biometrics Program as ACAT 1-Special Interest Program. Currently, no biometrics program of record (PoR) created by the EA exists for collection devices or for the authoritative database. U.S. Special Operations Command (SOCOM) has fielded a collection device under a PoR, and the Navy has a collection device PoR in development, known as the Identity Dominance System (IDS). All equipment currently deployed has been overwhelmingly supported by overseas contingency operations (OCO) funding.

---

[6] In 2004, Stenbit wrote, "To ensure that the mission needs and requirements for all IdM capabilities are adequately defined, an overarching vision must be developed. In memorandum of January 12, 2004, I established an Identity Management Senior Coordination Group (IdMSCG) as a DoD-wide focal point for policy, requirements, strategy, and oversight on physical and virtual IdM" (John Stenbit, "Development of an Identity Management (IdM) Vision for the Biometrics Program," memorandum, Washington, D.C., February 20, 2004b).

[7] Gordon England, Acting Deputy Secretary of Defense, "DoD Policy for Biometric Information for Access to U.S. Installations and Facilities in Iraq," memorandum, Washington, D.C., July 15, 2005.

[8] DoD, *Biometrics Enterprise Strategic Plan: 2008–2015*, August 27, 2008b.

[9] Use of biometrically based identification techniques is commonly divided into three categories: blue, gray, and red, each of which is described in more detail later in the document.

However, the era of large numbers of rapid acquisitions based on the Joint Urgent Operational Needs Statements (JUONS) and readily available OCO funding is coming to an end now that U.S. forces have left Iraq and are scheduled to leave Afghanistan after 2014. DoD must make difficult decisions about managing its biometrics "enterprise"—i.e., the collection of activities associated with biometrics collection and use—and the evolution of its biometrics capability.

Under the current official biometrics framework, DoD Directive (DoDD) 8521.01E, the EA is responsible for programming and budgeting for "sufficient resources to support common enterprise" aspects and coordinating "all component biometric requirements," among other things, while the DoD components are responsible for budgeting for their specific biometric needs.[10] In practice, through Acquisition Decision Memoranda (ADM), and based on other guidance, the Army has taken on roles in biometrics that go beyond DoDD 8521.01E. The Secretary of the Army was originally designated EA to "lead, consolidate, and coordinate all biometrics information assurance [IA] programs" of DoD by Public Law 106-246 before September 11, 2001, and the 2003 invasion of Iraq, when the context and reasoning for that designation was clearly different than today. Consequently, the changing budget and wartime environment prompted DoD to reexamine biometrics activities.

Study Purpose

In this context, the Principal Staff Assistant (PSA) for Biometrics within the Office of the Secretary of Defense (OSD) asked RAND to study the biometrics enterprise. Specifically, the PSA asked RAND to do the following:

- Assess the assignments and arrangements of the EA with respect to effectiveness and efficiency of meeting user needs as required by DoDD 8521.01E
- Provide a status report on the biometrics program for the Secretary of Defense, also as required by DoDD 8521.01E
- Provide the PSA with options and recommendations for the biometrics enterprise to help future planning and developing input for the Fiscal Year 2014 (FY14) Program Objective Memorandum (POM).

We understood "assignments and arrangements" to be all management decisions made by the EA himself and all management and programmatic activities carried out by the EA's components.

[10] DoD, "Department of Defense Biometrics," DoD Directive 8521.01E, February 21, 2008a, at paragraph 5.12.6 and 5.12.8.

How the Report Is Organized

Our report is organized into five chapters. The second chapter describes the study methodology and analytical framework we used. The third chapter assesses the assignments and arrangements of the EA, including the management structure and the functioning of the biometrics cycle. The fourth chapter provides the status report on the overall biometrics enterprise. The fifth chapter provides conclusions, options, and recommendations. The appendixes provide a summary of the Biometrics Enterprise Strategic Plan goal, the interview protocol, a discussion of metrics, and a list of references.

CHAPTER TWO

Study Methodology and Analytical Framework

Study Methodology

Our report is based on information collected from September 2011 through January 2012. Thus, subsequent changes in circumstances may affect minor portions of the facts described, notably a recent decision by the Army to shift some of the EA responsibilities from one part of the Army to another and ongoing work to upgrade the authoritative database and handheld collection devices. Although none of the known potential changes affect our analysis, conclusions, and recommendations, they should be examined in subsequent annual assessments and status reports.

The biometrics community is large and diverse. It stretches from the providers of biometrics services (mostly within the EA), to its customers within the combatant commands (COCOMs). It also includes the military services, overseers within OSD and the Office of the President, interagency peers in the Departments of State (DOS), Justice (DOJ), and Homeland Security (DHS), the Intelligence Community (IC), and others. Our study focused on the assigned and unassigned (but still executed) responsibilities and activities of the EA meant to establish and maintain biometrics capabilities. We also examined activities of other DoD components supported by and affecting the EA. It should be noted that while we reviewed the analytical and intelligence portions of biometrics activities as necessary to assess the overall status of the DoD biometrics enterprise and the assignments and arrangements of the EA, we did not assess the quality or sufficiency of those activities.

Since our task related to assessing assignments and arrangements of the EA with respect to the effectiveness and efficiency of meeting end-user needs, we chose a research approach utilizing document review and interviews with key stakeholders.[1] We reviewed relevant policy documents including DoDDs, memos related to the biometrics program, acquisition documentation, and program briefings. We reviewed numerous reports regarding the development of a

[1] No single document proved sufficient for identifying end-user needs. Other documents, including the Biometrics Enterprise Strategic Plan and Biometric Concept of Operations, supported the analysis but did not highlight end-user needs per RAND's tasking.

DoD biometrics program, including those produced by the U.S. Government Accountability Office (GAO), the Institute for Defense Analyses (IDA), and the Defense Science Board.[2] We reviewed literature on the theory of biometric use for identity management, its potential effects on security, and evaluations of various aspects of the current DoD Biometrics Program. We also interviewed customers of, and participants in, DoD biometrics. These interviews focused primarily on guidance received, current activities, constraints to effective operation, views of ideal operations, and suggestions for improvement.

This research approach is well suited for a study that aims to evaluate a large, multi-component program that: (1) affects multiple actors with different perspectives, and (2) lacks quantitative data upon which to base an assessment. The document review and interviews provided a basis for our discussion of the role of the EA, our discussion of the DoD biometrics enterprise writ large, and our depiction of the "biometrics cycle," i.e., the end-to-end capability starting with the input of directing the collection of biometric information and ending with use of the collected information (see Chapter Three). The interviews enabled us to gain an in-depth understanding of the experiences, perceptions, and insights of individuals with a range of roles in and around the DoD biometrics enterprise, including an understanding of the links betwen the EA's activities and the biometrics cycle. While quantitative analysis including development costs, speed of deployment, and fielding levels could have added to the information collected through documents and interviews, data were generally unavailable and of limited value in measuring such items as the effect of the EA on the biometrics cycle, the status of the DoD biometrics enterprise, and how well the EA fulfills its responsibilities in meeting end-user needs. The qualitative information obtained from the various sources was aggregated, checked for internal consistency, and validated if incompatible with comparable material. This provided the basis for our analysis of the DoD biometrics enterprise and the assignments and arrangements of the EA.

Interviews

Interviews are particularly useful for gathering detail on processes and systems and integrating the viewpoints of multiple individuals. Unlike surveys, interviews allow researchers to tailor questions for relevance and delve deeply into topics with each respondent. Through the interview process, we obtained detailed explanations of organizations' roles in, expectations for, and perceived risks to the biometrics cycle. By gathering many, varied perspectives, we could

---

[2] U.S. Government Accountability Office, *Defense Management: DoD Needs to Establish Clear Goals and Objectives, Guidance and a Designated Budget to Manage Its Biometrics Activities*, Washington, D.C., GAO-08-1065, September 2008a; GAO, *Defense Biometrics: DoD Can Establish More Guidance for Biometrics Collection and Explore Broader Data Sharing*, Washington, D.C., GAO-09-49, October 2008b; GAO, *Defense Biometrics: DoD Can Better Conform to Standards and Share Biometric Information with Federal Agencies*, Washington, D.C., GAO-11-276, March 2011; GAO, *Defense Biometrics: Additional Training for Leaders and More Timely Transmission of Data Could Enhance the Use of Biometrics in Afghanistan*, Washington, D.C., GAO-12-442, April 2012; Terry C. Heuring et al., *The Department of Defense Biometric Enterprise*, Alexandria, Va., Institute for Defense Analysis, IDA Paper P-4416, December 2008; Office of the Under Secretary of Defense for Acquisition Technology and Logistics, *Defense Biometrics*, Washington, D.C.: Department of Defense, March 2007.

develop a full, multi-layered description of the current status of DoD biometrics and its management from those engaged in the enterprise.

We developed a semi-structured interview protocol that was administered at each interview (see Appendix B). Utilizing such a protocol—instead of a structured one—allowed flexibility to explore unexpected or discrete topics in greater depth. The interview began with a description of the project, the sponsor, and RAND. RAND offered individual-level anonymity to interviewees through standard human subjects' protection protocols; RAND chose such an approach to ensure research integrity and to promote frank discussion on topics deemed highly contentious by the researchers, particularly questions regarding how well needs have been met. Questions focused on the interviewees' current roles and responsibilities involving biometrics, their organizations' uses of and requirements for biometrics and biometrics technology, how requirements are developed, how well operational needs are met, how gaps might be addressed, and recommended changes. Interviews typically lasted approximately 45 minutes. Interview notes were assembled for reference during the analysis.

We classified interviewees by role in the DoD biometrics enterprise to ensure adequate coverage of divergent participant perspectives. Our goal was to speak with members of the EA, Joint Staff, interagency partners, users of DoD biometrics, and representatives from DoD policy organizations, including

- Policy Organizations:
  - Office of the Secretary of Defense–Cost Assessment and Program Evaluation (OSD-CAPE)
  - Office of the Secretary of Defense–Chief Information Officer (OSD-CIO)
  - Office of the Under Secretary of Defense for Intelligence (OUSD[I])
  - Office of the Under Secretary of Defense for Policy (OUSD[P])
  - Office of the Under Secretary of Defense for Personnel and Readiness (OUSD[P&R])
  - Defense Intelligence Agency (DIA)

- User Organizations:
  - U.S. Africa Command (AFRICOM)
  - U.S. Central Command (CENTCOM)
  - U.S. European Command (EUCOM)
  - U.S. Pacific Command (PACOM)
  - U.S. Northern Command (NORTHCOM)
  - U.S. Special Operations Command (SOCOM)
  - Air Force
  - Navy
  - Marines
  - Coast Guard
  - Defense Manpower Data Center (DMDC)

- Joint Staff Organizations:
  - Joint Staff Force Structure, Resources, and Assessment Directorate (J8)
  - Joint Staff Deputy Directorate for Antiterrorism/Homeland Defense (J34)

- EA Organizations:
  - Army Deputy Chief of Staff for Intelligence (G-2)
  - Army Deputy Chief of Staff for Operations, Plans, and Training (G-3/5/7)
  - Army Deputy Chief of Staff for Resource Management (G-8)
  - Biometrics and Identity Management Agency–Crystal City (BIMA-CC)
  - Biometrics and Identity Management Agency–West Virginia (BIMA-WV)
  - Program Manager for Biometrics (PM-Biometrics)
  - U.S. Training and Doctrine Command (TRADOC) Capability Manager for Biometrics and Forensics (TCM-B&F)
  - Intelligence and Security Command (INSCOM)

- Interagency Organizations:
  - National Security Staff (NSS)
  - Office of Science and Technology Policy
  - Justice Department–Federal Bureau of Investigation–Criminal Justice Information Services (DOJ-FBI-CJIS)
  - Homeland Security Department–U.S. Visitor and Immigrant Status Indicator Technology (DHS-US-VISIT).

The initial list of interview candidates came from the Defense Biometric Contact List provided to RAND by the sponsor's office. Initial candidates were sent an email invitation from the RAND Principal Investigator, inviting them to schedule an interview with the project team members. Project members also traveled to the 2011 Biometric Consortium Conference (held that September in Tampa, Florida), in order to facilitate interviews with participants from diverse locations. In addition to candidates nominated by the sponsor, general announcements of the study were made at the 2011 Biometric Consortium Conference and at meetings with DoD biometric stakeholders throughout the study field period. Interested parties were asked to contact the RAND study team to participate in the research. Finally, during the interviews, participants were asked to identify additional individuals who should be invited to participate in the study. RAND team members followed up with suggested candidates to invite them to participate.

Our goal in selecting respondents was to include individuals who represented a range of roles and were informed about the DoD biometrics enterprise. This type of purposive sampling is meant to obtain data from individuals who could both speak with authority on the subject and were likely to hold diverse positions within the enterprise. More than 50 interviews were conducted across DoD and the U.S. government biometrics community. Although many other individuals undoubtedly could have met the requirements to participate, the interviews provided rich accounts of the status of the biometrics enterprise.

Analytical Framework

Data from the interviews and document review were used to assess the assignments and arrangements of the EA in meeting end-user needs and also to examine the overall status of the DoD Biometrics Program. The biometrics literature and policy documents allowed us to ascertain some responsibilities of the EA, understand the purported tasks in the biometrics cycle, and start to establish boundaries for the analysis within DoD biometrics activities. However, the documents did not provide sufficient guidance with which to complete our analysis. Instead, working from documents, supplemented by discussions with the PSA, we determined that the most useful framework for assessing the EA's assignments and arrangements was a description of the steps required to complete the DoD biometrics cycle in support of the end-user. We further determined that the most useful framework for completing a status report on the biometrics enterprise was to identify gaps in policy and implementation measures that could be inhibiting the enterprise. Information gathered during the interviews further shaped our understanding of the biometrics cycle and helped us identify issues within the EA, potential risks and inefficiencies in the biometrics cycle, and gaps in policy and implementation of biometrics-related efforts.

We developed our criteria for the assessment and status report to move from the prevailing approach of measuring program activity to focus more on user satisfaction and outcomes. While several data points about different biometrics activities are available, they do not speak to the user as an end goal, nor do they provide "counterfactuals"—program evaluation parlance for what would have happened given alternate arrangements. For example, BIMA noted that more than 900 matches had been made to the "High Value Individual Watch List" using biometric data. This is an important and valuable result of using biometric capabilities. However, we have no way to measure quantitatively how different assignments or arrangements of the EA would have affected the number of watch-list matches. We also have no benchmark against which to measure whether the 900 matches meet or exceed expectations. Similarly, we have no enterprise goals against which to measure the effectiveness of certain policies and implementation measures.

Logic Model Approach

We also considered a logic model approach to the assessment. According to one definition, "A logic model typically offers a simplified visual representation of the path of a program's operations, starting with inputs and then progressing to the program's activities, its outputs, its customers, and its intended outcomes. The model may also link the program's operations, what the program actually does either alone or with others to fulfill its mission, to its strategy."[3] Ideally, we would have been able to use a logic model to generate criteria by which to assess the EA's assignments and arrangements.

We analyzed the relevant DoDDs, biometrics program documentation, and interview information to understand the objectives of a biometrics EA and the purported progression from inputs of designated responsibilities to intended outcomes. Unfortunately, and surprisingly, we could not discern program objectives or trace a logical flow from inputs to outcomes in DoD biometrics because of a lack of consensus on priorities. Instead, we attempted to discern logical flows between the activities of the EA and those necessary to support the effectiveness and efficiency of the biometrics cycle.

Official documents show that the purpose of the EA shifted from creating a standard business process to be "pushed out" to the DoD components and incorporated in operations to the current biometrics program in which the EA is expected to respond to user requirements (i.e., "pull in" from users) with tactical devices supported by a common architecture.

The 2000 memorandum establishing a Biometrics EA clearly defines the programmatic objectives of a centralized organization to "ensure that biometrics technologies are integrated effectively into information assurance systems, physical access control systems, best business practices, and other DoD applications as appropriate."[4] However, a 2006 concept of operations (CONOP) for DoD biometrics shows the beginning of movement toward responding to warfighter needs. The Biometrics Task Force (BTF, now BIMA) formally recognized this transformed need in its FY07 Annual Report, writing, "As the focus of the BMO [Biometrics Management Office] and BFC [Biometrics Fusion Center] swung toward identification of enemy combatants and terrorists, Army and DoD leadership saw that biometric technologies were playing less of an IA-focused role and more of an operational role than originally envisioned."[5]

Shortly thereafter, the 2008 DoDD 8521.01E enshrined the EA's role by stating, "DoD biometrics programs shall be designed to improve the effectiveness and efficiency of biometrics activities throughout the Department of Defense by eliminating unwarranted duplication and overlap of technology development and information management efforts." With this statement, the main objective of the EA switched from pushing technology to pulling requirements that were simultaneously similar and incongruent, urgent and strategic, and pragmatic and developmental. The EA was also expected to determine how to improve the efficiency and effectiveness of these numerous activities. Yet, this requirements-based tasking preceded explicit definitions of what it means to improve efficiency and effectiveness. Further, the biometrics program's strategic plan contains goals that relate more closely to a technology "push"

---

3 Victoria A. Greenfield, Valerie L. Williams, and Elisa Eiseman, *Using Logic Models for Strategic Planning and Evaluation: Application to the National Center for Injury Prevention and Control*, Santa Monica, Calif.: RAND Corporation, TR-370-NCIPC, 2006.

4 De Leon, 2000.

5 U.S. Department of Defense Biometrics Task Force, *Annual Report*, 2007.

approach rather than a requirements "pull" approach, leading to confusion regarding the purpose of the centralized biometrics program.

In other words, while the EA is directed to support end-user requirements, guiding documents have neither clearly defined the end-users, nor established a process for prioritizing or standardizing requirements for multiple DoD components. The lack of agreement about whether the biometrics program is responsible to "push" capability or "pull" requirements, and to what end, complicates conducting an assessment.

The requirements process may be the factor most affected by the confusion about the centralized biometrics program's role. The process of determining what supports an efficient or effective DoD-wide biometrics program has remained imprecise and conflicting in the documentation. For example, BIMA's annual reports offer a map of its presumption about the biometric technology requirements process. However, it does not mirror the process that occurred in developing the PoRs for the handheld devices and the authoritative database. Further, as noted previously, the bulk of the biometric capability currently deployed resulted from urgent operational needs, whose requirements progressed through very different processes than the PoRs.

Assessment Approach

Consequently, we needed criteria for the assessment and status report that accommodated the diverging and shifting views of the biometrics program. We concluded that the most consistent representation of the intended outcome of current biometrics activities is the biometrics cycle. We proceeded methodically through the cycle to determine how the EA's activities affect the cycle. Thus, our criteria for the assessment of the EA's assignments and arrangements focus on potential risks posed to the biometrics cycle. Similarly, our criteria for the status report on the biometrics program focus on the potential risks posed to maintaining a biometrics capability in the future.

Based on our conclusions of this analysis, we offer recommendations that we believe will align future management, activities, and documentation with the goal of supporting end-users. We recognize the potential for numerous solutions to the lack of consensus for the program, and we offer one possible set of solutions.

CHAPTER THREE

Assessment of the Assignments and Arrangements of the Executive Agent

This chapter begins with a description of the EA organization, roles, and responsibilities. It then describes the biometric capabilities that fall under the purview of the EA. The third section assesses the assignments and arrangements of the EA and describes issues with each of the eight stages of the biometrics cycle and qualitatively assesses the risk that certain identified issues pose to the cycle.

The EA is supporting a functioning biometrics cycle that meets some level of warfighter needs and has had significant success identifying bad actors. Aspects of the EA's assignments and arrangements show some deficiencies in managing biometrics responsibilities and pose risks to the biometrics cycle and ability to provide for user needs. However, this assessment examines ten years of activity and was conducted at a single point in time. Thus, it does not fully capture the recent efforts by the EA to improve operations, many of which were ongoing during our research. The EA has worked to increase automation, formalize activities, and develop metrics that demonstrate the impact of biometrics work. RAND expects that the required future annual assessments will track these continued changes and efforts to improve effectiveness and efficiency.

Assignments and Arrangements of the Executive Agent

The responsibilities and functions of SECARMY acting as the EA for DoD Biometrics are spelled out in DoDD 8521.01E, paragraph 5.12, a document released by DoD in 2008 to promote a joint and coordinated organizational structure for a DoD-wide biometric capability. DoDD 8521.01E lists discrete tasks the EA must accomplish, but, in keeping with standard DoD practice, the directive does not dictate how the Army must specifically organize its various biometrics functions to carry out its responsibilities.

The Biometrics Cycle

However, before describing the specifics of the EA's management and functions, it is useful to briefly summarize what the EA is currently executing. In general, the EA executes those functions needed to enable and support the biometrics cycle for red- and gray-force information.

Figure 3.1 describes the biometrics cycle as derived from the PSA's description, DoD documentation, and interviews.

The EA's role in enabling and supporting the cycle includes developing, procuring, and fielding biometric collection, storage, and analytical technology, operating the authoritative database for red and gray forces, promulgating standards for biometric data, and collecting common biometric requirements.[1]

For most biometric records, the cycle currently operates as follows: An individual is "enrolled" by collecting (ideally) ten fingerprints, iris scans, a facial photograph, and biographic information (name, height, weight, etc.); the data are sent to the Automated Biometric Identification System (ABIS) to determine if a biometric match can be made to existing records; a match/no match response is sent to a recipient who may or may not be the person who submitted the request.[2] Most fingerprint records submitted to ABIS can be resolved by automated analysis, but some (approximately 9.5 percent as of January 2012 according to BIMA) must be reviewed by a human forensic examiner. Submissions are prioritized based on urgency, with SOCOM inquiries receiving higher priority than detention center inquiries (as well as Navy maritime interdiction operations, and FBI Hostage Rescue Team operations). Results of biometric submissions are also analyzed in conjunction with other available information about the identified person. As of January 2012, ABIS contained 7.1 million records linked to 4.5 million unique identities, and ABIS was processing, on average, approximately 6,500 transactions per day.

In addition to this process, BIMA employs latent print examiners who try to match ABIS records to partial fingerprints collected from various sources, e.g., IEDs. Fingerprint records are also received from partner nations working in cooperation with the United States.

Figure 3.1
The Biometrics Cycle

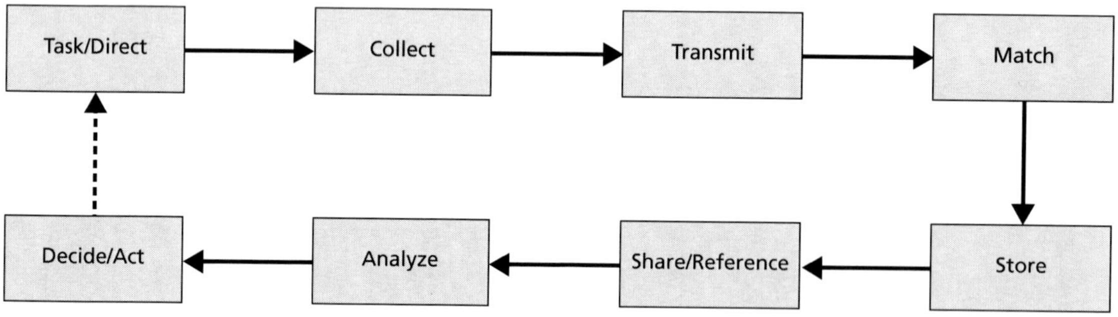

RAND *TR1290/1-3.1*

[1] The EA is also responsible for blue-force biometric work, but the current level of activity is much lower as described later in this report.

[2] One might suppose the cycle would terminate when a match/no match response is sent back to the field. It does not unless the match is to someone on the BEWL or the match request came from SOCOM. According to CENTCOM sources, the Project Manager for DoD Biometrics (PM-Biometrics) has had a requirement to implement a function in Biometrics Automated Toolkit (BAT) to be able to receive responses, but it has not been implemented.

EA Roles and Responsibilities

Prior to describing the biometrics-specific functions of the EA listed in DoDD 8521.01E, it is useful to reference DoDD 5101.1, which defines the general role of an EA within DoD. Conceptually, an EA is designated to ensure proper coordination among DoD components, identify requirements and resources, and monitor and report results of performance for capabilities not clearly within the purview of a single DoD component.[3] DoDD 5101.1 explicitly describes the role of an EA:

- "5.2.1. Execute DoD Executive Agent responsibilities, consistent with applicable law, DoD Directive 5100.3 (reference (d)), DoD Directive 5100.73 (reference (e)), and this Directive.
- "5.2.2. Ensure proper coordination with the DoD Components for the responsibilities and activities assigned to provide continuous, sustainable, and global support as required by end-users. Ensure effective planning throughout operations by developing a coordinated process and support plans for transition from peacetime to wartime and/or contingency operations.
- "5.2.3. Identify requirements and resources, including force structure to the extent permitted by law, necessary to execute assigned responsibilities and functions. Submit these requirements to the cognizant Head of the DoD Component to be included in their respective budget documentation.
- "5.2.4. Monitor resources used in performing assigned responsibilities and functions.
- "5.2.5. Develop, maintain, and report results of performance of DoD Executive Agent responsibilities and functions, as may be required by law, Secretary of Defense decision, or other Congressional requirements.
- "5.2.6. Obtain reports and information, consistent with DoD Directive 8910.1 (reference (f)), as necessary, to carry out assigned DoD Executive Agent responsibilities, functions, and authorities.
- "5.2.7. Establish, maintain, and preserve information as records, consistent with DoD Directive 5015.2 (reference (g)), that document the transaction of business and mission of the DoD Executive Agent.
- "5.2.8. Designate a focal point to coordinate matters regarding assigned DoD Executive Agent responsibilities, functions, and authorities."

DoDD 8521.01E specifies the role of the Biometrics EA in light of DoDD 5101.1 and the development and deployment of common biometric technology, operating the authoritative database, promulgation of data standards, and collection of requirements, among other things. In considering these duties, it is important to remember two things: First, the assignments and arrangements of an EA in general are meant to be monitored regularly for effectiveness in meeting end-user requirements. Second, DoDD 8521.01E preceded formalized strategy for a

---

3 DoD, "DoD Executive Agent," DoD Directive 5101.1, May 9, 2003.

coordinated DoD biometric capability and thus is merely an initial step. In light of these caveats, DoDD 8521.01E by itself does not (and cannot) serve as the basis for assessing the EA's assignments and arrangements. Instead, the directive is a framework for the entire biometrics program, including the EA's role, that is meant to be further specified by other implementation measures such as a DoD Instruction (DoDI). DoDD 8521.01E lists the explicit responsibilities of the Biometrics EA:

- "5.12.1. Execute responsibilities of the DoD EA for DoD Biometrics in accordance with Reference (w) and this Directive.
- "5.12.9. Develop, publish, and update as appropriate a DoD Biometrics Security Classification Guide.
- "5.12.8. Coordinate all component biometric requirements with DoD Component members of the DoD Biometrics EXCOM.
- "5.12.6. Program for and budget sufficient resources to support common enterprise requirements documentation, architecture development, materiel development, test and evaluation, lifecycle management, prototyping, exercises, records management, demonstrations, and evaluations to include efforts at maturing viable technologies and standards.
- "5.12.7. Program for and budget sufficient resources to support common biometric data management, training, operations, and lifecycle support.
- "5.12.3. Provide for, manage, and maintain a biometrics center of excellence.
- "5.12.2. Appoint an Executive Manager for DoD Biometrics, who shall be a G/FO or SES equivalent, with responsibilities as outlined in Enclosure 4.
- "5.12.4. Appoint a single Program Management Office, under the authority of the Army Acquisition Executive, responsible for the development, acquisition, and fielding of common biometrics enterprise systems to support common, Service, and joint requirements."[4]

We gathered additional information through interviews and document reviews to gain a more complete understanding of the EA's activities.

First, the EA appoints the Executive Manager for DoD Biometrics, who is currently the Director of BIMA. The Executive Manager is responsible for biometrics standards and promulgation, among other things. BIMA is an agency created by the Army responsible for carrying out the Executive Manager's functions, including managing the "authoritative DoD repositories of biometric samples on those individuals not issued a DoD credential."[5]

Second, the EA provides for, manages, and maintains "a biometrics center of excellence."[6]

---

[4] DoD, 2008s.

[5] DoD, 2008a, paragraph E4.11.

[6] DoD, 2008a, paragraph 5.12.3.

Third, the EA appoints "a single Program Management Office" for biometrics, currently PM-Biometrics, which is "responsible for the development, acquisition, and fielding of common biometrics enterprise systems to support common, Service, and joint requirements."[7]

Fourth, the EA, "when applicable ... make[s] recommendations to [the Under Secretary of Defense for Acquisition, Technology and Logistics] USD(AT&L) concerning acquisition category and milestone decisions for all biometric acquisition programs."[8]

Fifth, the EA is tasked with programming and budgeting resources to support common requirements. Currently, and over the last ten years, these resources have come overwhelmingly from OCO funding. This funding source relieved the Army of having to make trade-offs in its base budget between DoD-wide biometrics support and other Army needs. However, once OCO funding ends—a factor already starting to impinge on the FY13 program—biometrics will have to compete for funding and will therefore need a strong rationale in favor of surviving as an enduring DoD capability in some form. Furthermore, without OCO funding, biometrics requirements must be supported with formal PoRs created by the Army. As of this writing, the Army has initiated this effort, but it has moved only through Army's internal staffing process and must still complete joint staffing.

Sixth, the EA is responsible for programming and budgeting "sufficient resources to support common biometric data management, training, operations, and lifecycle support." Training has been ad hoc so far because formal training cannot be developed without PoRs.

Seventh, the EA is responsible for coordinating "all component biometric requirements with DoD Component members" of the DoD Biometrics Executive Committee (EXCOM), and, by extension, using these requirements in developing its acquisitions.

Finally, the EA is responsible for developing, publishing, and updating "a DoD Biometrics Security Classification Guide."

The EA has taken on and engaged in additional biometrics activities not specified in DoDD 8521.01E by virtue of additional OSD guidance, past practice, and the consequences of deployed systems. We identified these activities during the study and incorporated them into our understanding of the assignments and arrangements as appropriate.

The EA's management structure described by DoDD 8521.01E, officially formed by the EA, guided by ADMs, and as carried out in practice, is shown in Figure 3.2. Beyond the DoDD-required Executive Manager (BIMA) and PM-Biometrics, the management structure includes requirements and (in the future) training functions formally assigned to U.S. Army TRADOC, as well as BEI functions carried out by the National Ground Intelligence Center (NGIC). The EA components' primary functions and responsibilities as identified during this study are summarized in Table 3.1. For purposes of this report, we assumed all other EA functions are retained by the EA.

As described by the DoDDs discussed above, a PSA—currently the Assistant Secretary of Defense for Research and Engineering—oversees the EA's activities. For biometrics, the PSA

[7] DoD, 2008a, paragraph 5.12.4.

[8] DoD, 2008a, paragraph 5.12.5.

Figure 3.2
Executive Agent Management Structure

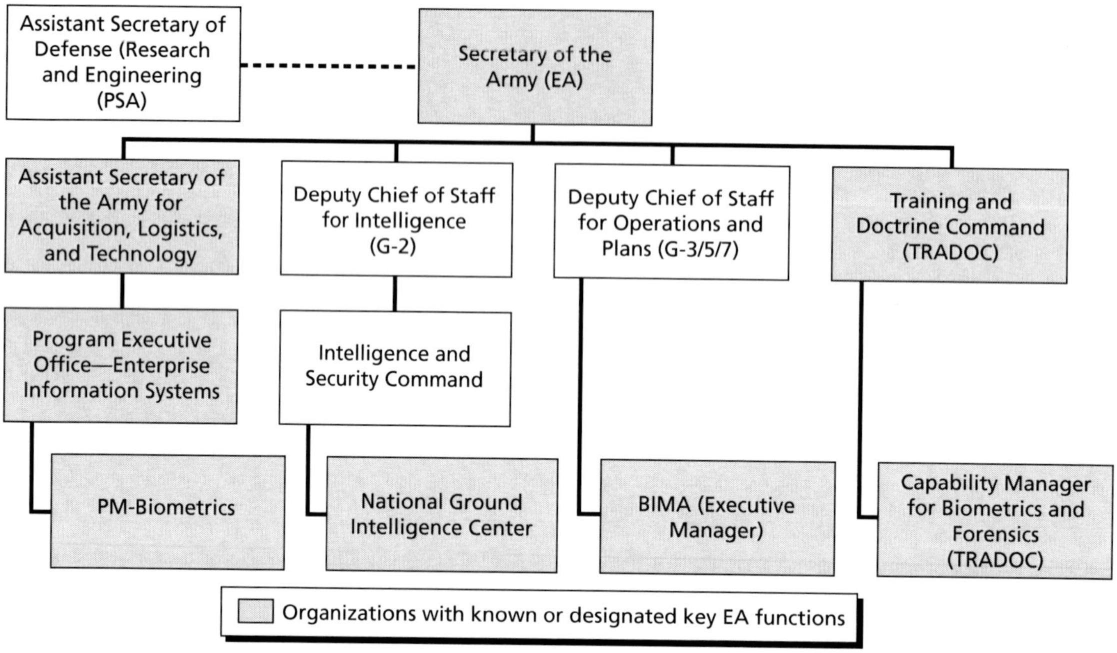

RAND *TR1290/1-3.2*

accomplishes his oversight primarily by chairing the EXCOM, which is the coordinating body for DoD biometrics. The EA is represented at EXCOM meetings by the Executive Manager, and the EA components provide the PSA with updates on biometrics activities. Other DoD components also participate in the EXCOM as needed.

Shortly after the publication of DoDD 8521.01E in 2008, DoD released the Biometrics Enterprise Strategic Plan (BESP). This document firmly prioritizes support of military operations for DoD biometrics, but, as discussed in Chapter Four, its goals are vague. As discussed in Chapter Two, when we tried to apply a logic model approach to track the program inputs to intended outcomes, we found the documents were inconsistent among themselves. This led us to use the approach that focuses on the biometrics cycle.

Assessment of EA Management Assignments and Arrangements

Several aspects of the EA's assignments and arrangements for carrying out its biometrics responsibilities are stovepiped, unclear, and duplicative, which could contribute to reduced effectiveness and efficiency of biometrics capabilities for end-users. RAND identified three factors as the likely root causes of management deficiencies:

- EA responsibilities spread across Army in a somewhat nontraditional manner

- continued reliance on OCO funding and quick reaction capabilities (QRCs)
- lack of objectives and metrics.

First, as noted above, the EA's functions—both those formally assigned and others taken on—are spread across the Army and only meet organizationally at the EA himself, SECARMY. The management arrangement is mostly typical within the Army, with separate roles for collecting requirements, procuring technology, and operating systems. However, the complicating factors in this case are the EA's Executive Manager and the need for centralized operations. Normally, requirements are gathered, equipment is acquired and fielded, and Army units use the equipment independently. With biometrics, the equipment must be used in concert with a central database that supports all of DoD. The EA's Executive Manager operates the database and has assumed some roles in technology development and requirements. Further, the EA's Executive Manager only has authority over BIMA, with no ability to coordinate the EA's functions and components.

Consequently, the closest approximation to the true role of Executive Manager has been taken on more by the PSA's representative, the Director of Defense Biometrics. However, as an

Table 3.1
Summary of Executive Agent Component Primary Functions and Responsibilities Identified in RAND Research

EA Component	Functions	Notes
BIMA – Executive Manager	• Vice chair of Biometrics EXCOM • Manage database for red and gray forces • Technology "support" • Standards • Develop tools to facilitate interoperability • Coordinate on continuity of operations (COOP) site	Responsible for all Executive Manager responsibilities listed in DoDD 8521.01E
PM-Biometrics	• Develop, acquire, and field biometrics systems • Database COOP site	Reports to Program Executive Office and Asst. Sec. Army on acquisition matters
TRADOC	• Gather biometrics requirements through TCM-B&F • Develop Capability Development Documents (CDDs) for PoRs through TCM-B&F • Develop and conduct training	
NGIC	• Operate BEWL • Provide BEI through analysts funded by the Military Intelligence Program (MIP) and National Intelligence Program (NIP) • Handle biometric data	BEWL and BEI functions directed to DIA in DoDD 8521.01E, but Army NGIC has been carrying them out

SOURCE: DoD, 2008a; and interviews with U.S. government personnel.

OSD office responsible for oversight and guidance, the PSA lacks directive authority over the EA's components. According to BIMA, "The EA's responsibilities in DoDD 8521.01E include additional items beyond what is included in DoDD 5101.1, thus distorting the perception of who has the responsibility to lead the DoD Biometrics Program. By allocating additional responsibilities to the PSA ... the EA focal point [in this case the Executive Manager] does not have the authority and responsibilities to adequately manage the DoD Biometrics Program."[9] We do not believe "additional" PSA responsibilities are hindering the Executive Manager, but the practical effect is that SECARMY, as the EA, remains the single point of management at a level in the Army that does not have the time to devote to the day-to-day minutiae of biometrics activities. The impact of this organizational positioning is shown in several instances. All non-Army personnel interviewed by RAND were unsure or unaware of which EA components are responsible for the EA's functions, and in fact, several Army personnel were also unsure about roles within the EA.

Because of its limited organizational reach, BIMA is unable to properly promulgate and enforce standards for biometric technology. Under DoDD 8521.01E, the DoD components are responsible for complying with the standards, and BIMA stated that the Biometric Standards Working Group, which BIMA chairs, has formally recommended standards to the DoD Information Technology Standards and Profile Registry (DISR) regularly since 2006. However, BIMA has not enforced the data standards, and technology has been procured that does not meet the standards.

Because of unclear responsibilities or unclear management direction, BIMA and TCM-B&F both ran requirements working groups, despite this function normally residing with a TCM. BIMA attempts to differentiate its working group by stating that it focuses on "collaboration among the services and federal partners," but TCM also works to include the other military services in the course of validating requirements.[10] Both TCM-B&F and G-3 said that having two groups operating in the requirements space is duplicative and inefficient. This duplication of effort—and potential lack of clarity about who was ultimately responsible—may have contributed to the approximately three years required to develop and complete the CDDs for BEC and Joint Personnel Identification version 2 (JPIv2), discussed in greater detail below.

A lack of collaboration between EA components has led to direct failures. PM-Biometrics and BIMA are both responsible for ABIS; BIMA for day-to-day management of operations and PM-Biometrics for fielding, upgrades, and maintenance. However, the organizations blamed each other for multiple failed ABIS upgrades—the most recent occurring in August 2011, with each group claiming the other failed to provide adequate information for the upgrade. Because of the stovepiped management structure, the conflict could only be resolved by the EA himself. Similarly, both PM-Biometrics and BIMA play a role in the COOP site for ABIS. DoD and

---

9 Comments from BIMA to RAND on the draft of this report, April 2012; DoDD 5101.1 lays out the general framework and scope of responsibilities for EAs, while DoDD 8521.01E provides the specific responsibilities and functions for the Biometrics EA.

10 Biometrics and Identity Management Agency comments to RAND, April 2012.

Army policy require COOP contingencies, and the EA has had a standing request for an ABIS COOP site since approximately 2007. As of this report, the COOP site is scheduled to be running in the first quarter of FY13, five years after the requirement was issued. At least one failed attempt to set up the COOP site has been made on the West Coast since 2007. PM-Biometrics stated, "It is the PM's job to establish a COOP site, but the PM has been unable to accomplish this mission because of lack of concurrence from BIMA."[11]

Second, most of the biometrics activities have been supported by OCO funding and have focused on procuring QRCs because of wartime environment demands. Much of the Army has been forced to respond to urgent operational needs from Iraq and Afghanistan, which resulted in tools and technology being rapidly developed and fielded without adhering to DoD standards, formal performance measures, and operational testing and evaluation requirements. This was seen as important for early fielding given the pressure of wartime. However, the biometrics enterprise continued operating under this paradigm, lacking firm deadlines and accountability measures, according to a G-3/5/7 interview. The absence of structure and strategic planning contributed to the proliferation of devices that are not completely interoperable and to conflicts between end-users and EA components over which tasks were given priority. For example, PM-Biometrics was criticized by part of the Army for prioritizing work on the Biometric Identification System for Access (BISA) rather than the BAT, which was in greater use at the time. PM-Biometrics claimed the apparent difference in effort was due to BISA's code having "greater maintainability and sustainability."[12]

Third, the EA lacks metrics to determine the sufficiency and efficiency of its activities. When RAND first requested performance metrics from the EA components, the responses largely consisted of budget execution (i.e., spending) and staffing levels. BIMA-WV explained how it tracks ABIS daily transactions, number of matches, and downtime, but these are metrics with mostly self-imposed (i.e., BIMA-imposed) thresholds and objectives and without a basis to determine sufficiency, i.e., whether warfighter needs are being adequately met.[13] As discussed below, DoD as a whole lacks goals for using biometrics and implementing biometric technology, but the EA is still responsible for developing organizational goals and objectives to provide guidance and enforce accountability. BIMA identified the DoD BESP 2008-2015, issued August 27, 2008, as a source of goals and objectives. However, the strategic plan is vague and generally tracks DoD's broad mission of supporting the warfighter. Also, the strategic plan was issued in 2008 while BIMA was still known as the BTF, U.S. troops were still present in Iraq, and no formal PoRs had been developed.

These root causes of management issues are inherently linked and manifest in different ways within the EA's components. PM-Biometrics could take any approach it deemed appropriate—and as approved by its immediate superior, Program Executive Office for Enterprise Information

---

[11] Comments from PM-Biometrics to RAND on the draft of this report, April 2012.

[12] PM-Biometrics comments, April 2012.

[13] BIMA comments, April 2012. BIMA noted that the threshold of the maximum time to respond to SOCOM inquiries was set by SOCOM.

Systems (PEO-EIS)—to address the urgent operational needs and requests for changes in biometrics equipment. However, these decisions essentially were made in a vacuum because there were no programmatic goals or metrics and no links between OCO funding and programmatic output. Further, the organizational stovepipes meant PM-Biometrics did not have to collaborate with BIMA in making decisions. According to CENTCOM, more than 80 percent of validated requirements submitted from theater to PM-Biometrics were not completed.

Despite these issues, the biometrics capability maintained by the EA has helped make more than 900 matches to the High Value Individual Watch List and deny more than 80,000 attempts by people to gain employment, access, or training at U.S. or coalition facilities.[14]

It is important to note that beginning in 2008, GAO wrote multiple reports highlighting some of these problems with management.[15] The EA also hired ANSER Analytic Services (and a team of study partners) to provide a series of reports.[16]

The EA is taking some initial steps to correct highlighted problems, such as imposing the structure and discipline of formal PoRs and moving ABIS operations from BIMA to another Army organization.[17] However, the three root causes—stovepiped management, OCO funding of QRCs, and a lack of objectives—remain essentially unchanged and therefore will likely continue causing difficulties. How the EA manages biometrics responsibilities through the various components is discussed in greater detail in the following section.

EA Management: BIMA and PM-Biometrics

BIMA (and predecessor organizations) and PM-Biometrics have played the most prominent roles for the EA because they have specifically designated roles and are collectively responsible for most of the biometrics technology and data.

BIMA and predecessor organizations have had significant successes in supporting warfighter needs for biometrics matching, but they have struggled to define themselves over approximately 12 years. Responsibilities were initially split between the BMO and BFC under G-6 in July 2000. The organizations were merged into the BTF under G-3/5/7 in June 2006, which was then changed to BIMA in March 2010.[18] BIMA's responsibilities are being reorganized again as of the end of our research effort, and BIMA's functions will be made subordinate to the Office of the Provost Marshal General by the end of FY12. This instability of role and management has likely been at the root of some of BIMA's (and predecessor organizations') internal management deficiencies described here. BIMA has mostly self-imposed performance objectives for operating ABIS, its most important function, and no organizational metrics or goals that were shared with

---

14 BIMA comments, April 2012.

15 GAO, 2008a; GAO, 2008b; GAO, 2011; GAO, 2012.

16 See, e.g., Team ANSER, 2009a, 2009b, 2009c.

17 John M. McHugh, Secretary of the Army, "Future of the Biometrics Identity Management Agency," memorandum, Washington, D.C., March 12, 2012.

18 At least one interviewee stated that G-2 had responsibility for part of the biometrics mission between G-6 and G-3/5/7. This was denied by BIMA but is a further indication of the lack of clarity about roles and responsibilities.

RAND. BIMA notes that it has a threshold requirement of 8,000 transactions per day for ABIS and that actual performance recently increased to 18,500 per day. This self-imposed threshold is easily achieved, given current performance, and is not based on a DoD-wide (and potentially U.S. government-wide) understanding of the true need, which could be higher or lower. BIMA has spent millions of dollars on multiple pilot efforts without a transition plan for any technology it develops, and without any successful transition of technology to DoD users. Exact figures were unavailable, but in recent years, these efforts have averaged in excess of $3 million per year. As one BIMA staff member indicated, BIMA in recent years has self-initiated research efforts without a specific request or an obvious need from an end-user.[19]

BIMA's inability to effectively promulgate and enforce standards (discussed in greater detail below) for biometric data collection contributed to the multiple devices procured across DoD that are not fully interoperable and, in the case of the Handheld Interagency Identity Detection Equipment (HIIDE), did not record data that meet the FBI's fingerprint standard. BIMA has taken steps to improve procedures for ensuring that biometrics equipment conforms to standards by obtaining accreditation from the National Institute of Standards and Technology (NIST) under the National Voluntary Laboratory Accreditation Program. However, as highlighted by GAO, "the current DODD 8521.01E already requires such compliance testing for new biometrics acquisitions, but DOD noted and we [GAO] agree that the directive does not fully address quick reaction capabilities such as the Handheld Interagency Identity Detection Equipment (HIIDE)."[20] Thus, the first significant opportunities to prove the value of these efforts will not come until new equipment is under development through the new PoRs.

In contrast to BIMA, PM-Biometrics has a relatively clear set of responsibilities for procuring and developing biometrics technology, but the execution of these responsibilities lacks any measures of sufficiency or efficiency and has been criticized by biometrics users in the field, among others. The most frequent complaint about PM-Biometrics from interviewees was that it did not work on the top user priorities, focusing efforts instead on priorities of its own choosing. CENTCOM noted that PM-Biometrics has not completed the four JUONS submitted since 2009, with only the replacement for the HIIDE nearing completion. CENTCOM also stated that PM-Biometrics did not complete "89% of Army [Department of Army] DA G3 Validated Requirements in over 4[-plus] years."[21] Assertions about PM-Biometrics's approach are further supported to some extent by the fact that the only performance metrics PM-Biometrics initially provided RAND were numbers of staff and funds expended.

PM-Biometrics' response to this criticism, in part, was to point out that it does not have formal requirements. However, PM-Biometrics noted that its work priorities are captured in Configuration Change Boards. Also, as noted above, the lack of goals and objectives is a deficiency throughout the EA and all of DoD. This is highlighted by recent delays and changes in schedules and funding. Between December 2011 and January 2012, PM-Biometrics pushed

[19] Interview with BIMA staff member, September 2011.

[20] GAO-11-276, March 2011.

[21] Comments from CENTCOM to RAND on the draft of this report, May 2012.

back the estimated date to replace the HIIDE devices with Secure Electronic Enrollment Kit version II (SEEK II) devices by at least three months.[22] PM-Biometrics explained that U.S. Forces Afghanistan (USFOR-A) had added a requirement that the new SEEK II needed to be able to communicate through the BAT infrastructure, which, according to PM-Biometrics, "is a step backwards for the SEEK as it perpetuates the BAT-[Army version 4]Av4 inefficient architecture." However, PM-Biometrics's original plan for the SEEK appeared to rely on BISA, which is being phased out. Further, between November 2011 and January 2012, BEC funding for FY13 through FY16 was decreased from $35 million per year to $3 million per year.[23] Without formal PoRs that impose deadlines and accountability measures, PM-Biometrics appears to some users to be setting its own priorities (with approval from PEO-EIS), while PM-Biometrics complains of being forced to maintain existing inefficient technology. Neither approach is based on true measures of performance and effectiveness.

Moreover, DoD lacks a process for prioritizing JUONS, so PM-Biometrics may be receiving multiple requests that are competing for limited resources. According to G-3, the Army has put in place a process to track JUONS and the responses to them, but DoD as a whole lacks a process to prioritize JUONS. BIMA stated that "Part of the issue also resides with the [Joint Rapid Acquisition Cell] JRAC for approving and validating JUONS that were not feasible ([e.g.,] JUONS 0434), instead of requesting a capability and describing a problem the JUONS has [sic] approved as [a] wish list of solutions."[24] However, this only partly explains CENTCOM's statement that PM-Biometrics has not completed the four validated JUONS submitted by CENTCOM since 2009.

EA Management: TRADOC and NGIC

To date, TRADOC's most important functions related to biometrics have been to assign TCM-B&F responsibility for gathering requirements and developing the CDDs for the BEC and JPIv2 PoRs. As noted above, the CDDs were approximately three years in the making (the delays attributed to various and conflicting reasons), but G-3/5/7 complimented TCM-B&F's work and inclusiveness in collecting and documenting the requirements.

TRADOC's most notable deficiency is the fact that no formal biometrics training program can exist for equipment that did not originate from a PoR. Training remains ad hoc despite biometrics technology being fielded and used for almost ten years. TCM-B&F, which is not responsible for the eventual formal training program, has taken steps to develop biometrics training simulations, but even these are not being provided in sufficient quantity for pre-deployment training. However, BIMA claims TCM-B&F "continue to do development associated with BAT, [Multilingual Automated Response System] MARS, et al [sic] despite not having acquisition authority in biometrics."[25]

---

[22] PM-Biometrics comments, April 2012.

[23] Comments from Army personnel to RAND on the draft of this report, January 2012.

[24] BIMA comments, April 2012.

[25] BIMA comments, April 2012.

As noted above, NGIC is not formally called out as part of the EA, but it has assumed some EA functions as well as some of DIA's assigned functions for biometrics. NGIC has been acting as an intermediary between the field and ABIS for matching requests and has provided BEI for all DoD components through MIP- and NIP-funded analysts and resources. In late 2011, Army announced it would no longer provide analytical support to non-land forces and potentially all non-Army organizations (as of February 2012, G-2 and DIA were staffing and discussing possible options). This uncertainty creates difficulty for non-Army component planning. Moreover, Army is reluctant to relinquish control over the NIP-funded resources despite not having official control over them. These issues need to be resolved to ensure sufficient analytical resources are available for non–Army-specific missions going forward.

Current Biometrics Capability Under EA Purview

The biometrics capabilities currently under the EA purview comprise a variety of technological solutions that arose by matching available technology to operational needs on a quick reaction basis. No structured, programmed capabilities existed prior to OCOs in Iraq and Afghanistan, and DoD biometrics priorities before that did not include tactical collection. The operational demands of two wars required the EA to begin providing tactical equipment, which diverged from the EA's original assignments that were based on more hypothetical ambitions of DoD biometrics. This divergence was not fully resolved in DoDD 8521.01E. In an unintended fashion, the tactical collection devices—which were not part of any articulated vision or procurement strategy of the DoD biometrics community—have become the foundation of DoD's biometrics collection capability.

The limitations of the technology currently under the EA purview are well known, particularly in the PoR requests for the BEC and JPIv2. The GAO reports about the DoD Biometrics Program also highlighted deficiencies in the enterprise.[26] We do not attempt in this report to characterize specific root causes for technological constraints of each technology; instead, we offer two broad conclusions regarding the impact of utilizing QRC acquisition for biometrics equipment. First, as noted by U.S. Joint Forces Command, and GAO, "Without a formal program for biometrics, not all steps associated with safeguards in DoD's acquisition process for new technological systems are occurring."[27] Second, QRCs force trade-offs between capabilities currently available on the market and long-term needs. In other words, DoD accepted capabilities with some limitations because they were readily available commercially and could be fielded quickly. This report does not suggest that using QRCs was inappropriate; in fact, it recognizes their benefit in providing the warfighter with tactical collection devices whose need was unforeseen by strategic planning.

---

[26] GAO, 2008a; GAO, 2011.

[27] GAO, 2008a.

We know efforts are under way to update and improve the current systems, which makes writing an assessment and status report more difficult. We also know the EA supported a wide range of other "pilot" studies and research projects, but the information we received did not reveal any instances of those projects being fielded or transitioned to operations. Further, GAO noted in a 2008 report the inability to definitively track programs through spending, particularly short-term OCO-funded pilot programs.[28] Therefore, rather than writing about speculation or moving targets, we chose to discuss the systems that are currently fielded. It is worth noting that the EA is working on or supporting multiple pilot-scale efforts and studies, and these should be examined in subsequent annual assessments.

The technologies being used have three main functions: biometric data collection, data storage, and matching. The devices used by DoD include handheld collection devices, tabletop collection devices with local storage and matching capability, and the authoritative database for storing and matching data on red and gray forces.

As QRCs, the fielded systems do not have documented performance thresholds and objectives vetted by the Joint Requirements Oversight Council (JROC)—a process designed to ensure user requirements will be met. While we cannot comment on or report current device performance as measured against vetted user requirements, we can present current, fielded capability. While these capabilities may differ from those documented in evolving, draft CDDs and capability production documents (CPDs), we are providing an assessment and status report based on the biometrics program as it is, rather than as it is planned to be.

The EA is also developing two new PoRs: BEC Increment 1 for the authoritative database (replacing ABIS); and JPIv2 for mobile and handheld collection devices (replacing BAT, HIIDE, and SEEK II). The status and implications of the new PoRs are discussed later in this report.

Assessment of EA Assignments and Arrangements Supporting the Biometrics Cycle and End-User Needs

Apart from assessing the assignments and arrangements of the EA in terms of managing its responsibilities from a process perspective, it is also necessary to examine the assignments and arrangements in terms of support to warfighting needs. In other words, is the EA supporting a functioning process for collecting, transmitting, storing, matching, sharing, and analyzing biometrics?

The basic answer is yes. Biometrics information is being collected and has been used to identify individuals against prior encounters and forensic evidence. The authoritative database for red and gray forces is growing steadily, and the processes in the field are becoming increasingly routinized. In that sense, hard work and concerted efforts by many involved have produced a functioning biometrics cycle that is helping identify bad actors and verify the identities of other personnel.

---

[28] GAO, 2008a.

A deeper examination, however, suggests risks and room for improvement. To explicate as much, we use a heuristic metric that assumes the perfect biometrics cycle is one in which warfighters can collect accurate biometrics, have them analyzed within the limits set by technology (typically, in a few minutes), stored indefinitely and securely, and shared with others. Correspondingly, the actual biometrics cycle can be assessed in terms of risks to operation, actual impediments, and potential failure modes. Our analytical method, then, is to examine each stage of the cycle for potential impediments and risks. Some can be traced back to issues associated with the EA's management approach, but some of them are issues that may be influenced but not determined by the EA.

In the sections that follow, we discuss our assessment of each segment of the biometrics cycle, with a focus on issues that reduce or potentially reduce cycle efficiency and effectiveness. The tables presented with each section summarize the issues with that portion of the cycle, implications, status of the issues, and the reasons.

We attempted to differentiate between our assessments of high- and medium-risk issues (shown by "H" and "M"). High-risk issues are those already creating a risk of failure, or highly likely to do so in the future. Medium-risk issues are those that could pose a risk of failure in the future.

The goal is to highlight only those areas we assessed as posing some risk to the cycle. The discussion and tables that follow cover potential faults in the cycle. The tables highlight the link between the potential fault, the implications if a fault occurs, and the reason for the potential fault. We do not intend to portray the EA's assignments and arrangements as all negative. Rather, in the absence of overarching goals and objectives, we focused on areas where the EA could make changes to potentially improve how the cycle functions. We acknowledge the difference between "high" and "medium" risk is somewhat subjective, but we believe these are the issues posing the greatest risks to the biometrics cycle based on the data we collected.

Issues: Task/Direct

All biometrics work starts with a task or direction, either implicit or explicit, to collect the biometrics before transmitting or processing them. Hence the question: Under what circumstances should biometrics be collected?

Ordinarily, like most military taskings, this would be a decision best left to the local commander who has an operation to conduct and a set of resources with which to conduct it. The decision to collect biometrics would seem to fall into the category of unquestioned local control; time is a valuable resource, and time spent collecting biometrics cannot be spent otherwise.

Upon re-examination, however, the answer is less straightforward. People move from place to place, even from theater to theater. Al Qaeda, for its part, has a global reach, and someone encountered in Iraq may be re-encountered in Afghanistan, in PACOM's area of responsibility (AOR), or even attempting to enter the United States, and failure to collect biometrics in one location could prevent warfighters in another location from realizing that the person they are encountering has movements that may suggest problematic associations. The fact that information collected in one theater can influence operations in another suggests that some consistency in biometrics collection may be beneficial. (Whether it is beneficial enough to mandate a vigorous collection policy everywhere is another issue.)

Current red-force biometrics collection requirements are inconsistent across AORs. This is not the EA's fault inasmuch as combatant commanders have the authority to set local practices regarding such issues. However, it must be noted that the EA did not outline a clear direction to govern the collection of biometrics, and correspondingly, did not impress upon theater commanders some de minimus level of performance.

A second potential risk in the task-and-direct component of the biometrics cycle is a disconnect between the capabilities chosen for development and the capabilities that best reflect warfighter requirements. Our interviews indicated that although the EA is sensitive to Army requirements, it is less sensitive to requirements arising from the Marine Corps, Navy, CENTCOM, SOCOM, and the IC. The result has been that several of the communities, notably the Navy, have decided to look elsewhere to meet their needs.

A final risk is associated with the path biometric data takes from the war zone into the database. The Army has sent its material to NGIC and from there to the authoritative database ABIS, operated by BIMA-WV, where it then becomes available for others to use.[29] But NGIC has announced that it may restrict its services to Army customers alone, once biometrics activities are no longer funded under OCO accounts. Further, the flow of data has not been satisfactory to all users. CENTCOM "has had a requirement for four years to have an automatic guard in place to move biometrics records [from classified to unclassified systems for direct transmission to ABIS] with no success."[30] SOCOM has developed, and the Marines have pilot-tested, other pathways into ABIS. SOCOM's procedure involves satellites and web-based portals with direct priority access to ABIS and the latent print examiners at BIMA-WV, which has the advantage of guaranteeing a rapid response to their urgent requests. But allowing the various biometrics-collection communities to use different paths to the same goal has to come at the cost of efficiency. It is difficult to avoid the impression that the EA's priorities are tilted toward Army requirements even though the office holds a DoD-wide charter. Table 3.2 summarizes the issues with the Task/Direct portion of the cycle.

Issues: Collect

We identified several problems that may hobble the collection part of the biometrics cycle. Although none is a show-stopper individually, collectively they degrade the efficiency and effectiveness of the overall effort.

The first problem is that the various instruments currently in the field for collection—the HIIDE, BAT, and BISA—all produce non-standard data. SEEK II is expected to produce standard data, but as of this writing, none of these units has been fielded (except for SEEK

---

[29] CENTCOM, however, has not been satisfied with this arrangement. In 2008 it submitted a requirement to put an automated cross-domain solution (a classified-to-unclassified "automated guard") put in place to move biometrics—but with little success to date.

[30] CENTCOM comments, May 2012.

Table 3.2
Issues Associated with the Task/Direct Portion of the Biometrics Cycle

Cycle Issues	Implications	Risk	Reason
Inconsistent collection across AORs both in terms of whom to collect on and what data to collect (Collection plans are not an EA responsibility)	Bad actors detected in one theater may not be detected in another	M	DoD has not outlined clear guidance for collection in terms of targets and data to be collected
Capability development does not meet warfighter priorities relevant to circumstances	Devices not available for full range of potential missions	H	Lack of inter-military service requirements prioritization for QRCs Product development funding in PM's control, and PM priorities appear to differ from warfighter's[a]

[a] PM-Biometrics comments, April 2012. PM-Biometrics stated that "its priorities are completely in synch with the warfighters' ... what the warfighter fails to understand is that it is the PM's responsibility to provide the best solution to the problem and not just to provide what the warfighter requests."

devices procured separately by SOCOM).[31] There are two implications of collection devices not meeting standards: First, the data files' quality fail to meet minimum levels; and second, the data are not in the right format. In the first case, data could not be used everywhere. With fingerprints, the usual problem is that while the center of each finger is recorded (as it would be in flat or "slap" prints), the side of the finger is not (as it would be in rolled prints), and thus its usefulness for forensic investigation is substantially reduced. In the second case, data could not be interchanged between devices, or images produced by one device could not be used with the matching algorithm associated with the authoritative database. That noted, in practice, there have been few complaints about the inability to match individuals. However, it is possible that crimes and attacks (e.g., using IEDs) cannot be ascribed to an individual because the fingerprints left at the scene do not include enough of the finger.

The world of biometrics—which, among other tasks, converts observations of the real world into zeroes and ones—is consequently ringed with standards, and adhering to them is important in making the enterprise run smoothly. Standards, as a rule, cover two broad areas: performance and conformance. A performance standard for biometrics, for instance, is one that says that an acceptable fingerprint image provides at least 500 pixels to the inch; an image that boasts a sharper resolution would also meet standards.[32] A conformance standard specifies how something is expressed. The Electronic Biometric Transmission Specification (EBTS) suite of standards delineates how the information required to transfer biometric information is

[31] That noted, there are multiple software configurations and proprietary standards associated with processing data from different SEEK vendors.

[32] U.S. Department of Commerce, National Institute of Standards and Technology, *NIST Special Publication 500–271: Information Technology: American National Standards for Information Systems—Data Format for the Interchange of Fingerprint, Facial, & Other Biometric Information – Part 1*, ANSI/NIST-ITL 1-2007, Gaithersburg, Md.: Information Access Division, Information Technology Laboratory, National Institute of Standards and Technology, May 2007.

to be encoded (in Extensible Markup Language [XML]). A nonstandard encoding would be one in which the wrong words were used or the categories associated with a specific description (e.g., meaningful error conditions) are not identical with those of the standard (e.g., a term used in the submission overlaps in definition with parts of two terms in the standard). Here, the point is not to exceed the standard but to line up with it. But even conformance standards can have some performance issues; for example, a substandard record could be one in which information was not provided for required fields.

The EBTS standards in DoD have been a source of confusion and frustration. In December 2009, BIMA brought the issue of retiring DoD EBTS 1.2 before its Biometrics Standards Working Group (BSWG), which voted to mandate EBTS 2.0 and retire EBTS 1.2 in the DISR. The change technically became effective in March 2010. However, none of the fielded systems were EBTS 2.0 compliant, and other work was already in progress based on EBTS 1.2. After learning of the potential change to EBTS 2.0, the Navy requested implementation guidance in November 2009 for its IDS program, which was working toward a design review for March 2010. BIMA responded to that request in April 2010 by granting a waiver to EBTS 2.0, allowing the Navy to continue based on EBTS 1.2. Similarly, in May 2010, PM-Biometrics requested a waiver because the ABIS upgrade would not be able to accommodate EBTS 2.0 while the data being received from collection devices were still conforming with EBTS 1.2. According to the Navy, EBTS 2.0 implementation guidance had not been provided as of May 2012.

Deficiency in the collection devices also originates with the QRC process. QRCs are intended to prioritize equipment that can be rapidly fielded over the so-called 100 percent solution.[33] For example, the HIIDE procured as a QRC had limited capability to ensure speedy deployment. It was known at the time of acquisition that the HIIDE did not meet agreed-to EBTS interagency fingerprint-sharing standards, but the technology was the best that was commercially available and could be fielded quickly. The problem, our interviews found, was that the HIIDE continued to be purchased and fielded even after improved commercial off-the-shelf technology became available—namely the SEEK II, which meets interagency data-sharing standards. We are still unclear why the HIIDE QRC was retained with its substandard fingerprint component when an improved device could have been fielded sooner within the same QRC framework.[34]

A problem of greater consequence is that soldiers are not very proficient at data collection or entry, resulting in some number of biometrics records not meeting quality standards.[35] In some cases, the captured biometrics are not easily readable. In other cases, elementary mistakes are made and not caught, e.g., identifying the left hand's fingerprints as the right hand's and vice versa. While OCO funding was available, soldiers' shortcomings could be overcome by paying for contractors to take biometrics in the field, but this expedient did not create a legacy

[33] U.S. Army, *Army Regulation 71-9: Warfighting Capabilities Determination*, Washington, D.C.: U.S. Army, December 2009.

[34] Repeated JUONS indicated issues with HIIDE, including poor test performance. PM-Biometrics argues that it was able to purchase HIIDEs and not SEEKs because it had sole-source purchasing authority for one but not the other.

[35] BIMA maintains that there is no biometric image quality requirement for DoD collections.

base of knowledge to build upon, and, now that the Army has to fund biometrics from its own budget, is unlikely to be tomorrow's solution.

A cause of such shortcomings is the lack of biometrics education in the DoD schoolhouses. This was explained by the fact that Army doctrine does not permit formal training programs on devices that are not products of PoRs. Presumably, as the Army is rolling out the products of the PoRs (circa FY15), such gaps will be addressed. Nevertheless, the problem of inadequate training and education will continue, and as noted by GAO, "biometrics training for leaders does not provide detailed instructions on how to effectively use and manage biometrics collection tools."[36] The Army will require substantial time to fund, develop, and actually implement new training modules. Table 3.3 summarizes the issues with the "Collect" portion of the cycle.

Issues: Transmit

Neither Afghanistan nor Iraq (at least when U.S. forces were operating there) can be described as places with good wired or wireless infrastructure. Thus, transmitting information from the field to ABIS—something that would be easy in this country—instead became another risk in the cycle. Those who collect biometrics often must wait until the data captured in the handheld devices are transferred to BAT terminals wired to the SIPRNet. Such a delay is measured in hours, sometimes days, partly because the HIIDE cannot transmit data directly and partly because of the sparse wireless network and available bandwidth in theater. In theory, the BAT terminal can match data locally to avoid the need for ABIS-based responses for all inquiries, but according to PM-Biometrics, this can only happen for high-quality fingerprints compared

Table 3.3
Issues Associated with the Collect Portion of the Biometrics Cycle

Cycle Issues	Implications	Risk	Reason
HIIDE, BAT, and BISA produce nonstandard data	Inefficient sharing of data among mainstays of DoD biometrics collection and interagency partners	M	Standards not enforced with QRCs Excessive retention of legacy QRCs
HIIDE does not meet collection standards SEEKs, which meet more standards, planned for fielding in FY12	Uncertainty/failure-to-match rate greater than best practice and limits sharing with interagency partners[a]	M	Indefinite retention of legacy QRCs
Training is ad hoc Soldiers not proficient at data collection or entry	Inconsistent, inaccurate and unusable data	H	No PoR with corresponding training programs

[a] BIMA comments, April 2012. BIMA maintains that there are no defined best practices in regards to failure-to-match rates for handheld devices.

[36] GAO, 2012.

against equally high-quality fingerprints already stored with the subset of data on the BAT, referred to as "lights out" matches by PM-Biometrics staff.[37] In practice, if biometrics are loaded into a BAT device, sufficient SIPRNet bandwidth is available to conduct a matching inquiry through ABIS in West Virginia.

As noted, there are three entry points for biometrics: BISA, BAT (including data collected by the portable HIIDE device used by conventional forces), and the portable SEEK device used by SOCOM. If the biometrics are acquired by conventional forces in connected locations (such as detention centers), the prints are sent to NGIC to strip out biographical and contextual data on the individuals whose biometrics have been captured. Once such data are stripped, the unclassified biometric data are transferred from SIPRNet to NIPRNet by means of a dedicated cross-domain tool (also known as a network guard) and forwarded to ABIS for a biometrics match.

SOCOM's frustration with this process prompted it to acquire a capability to communicate from the field through a satellite link to a web-based portal that is accessed by BIMA on the other end, which allows SOCOM's match requests to get high priority.[38] The result seems to have satisfied SOCOM's requirements, despite the added cost of using satellite instead of landline bandwidth, as long as examiners at BIMA-WV remain available to give SOCOM top priority.

In late 2011, consistent with the Army's evolving approach to its joint requirements, NGIC announced that, as of March 1, 2012, it would phase out its services to all warfighting components apart from the Army's. Army has since begun considering a range of options for potentially supporting non-Army requirements. Whether this will create major delays for biometrics collected by the Marines and the Navy rests on how successfully and quickly ABIS can get its own SIPR-to-NIPR cross-domain solution operating. In November 2011, their expected go-live date was December 2011, but it slipped to April 2012. This remains a high warfighter priority, and the technology has existed for decades.

The EA acknowledged both its role in communications architecture and the limits of its ability to influence the field, which is mainly outside the EA's purview. The EA (through BIMA) provided approximately half the funding for the pilot study of the Last Tactical Mile (LTM), a CENTCOM-initiated JUONS to send biometric data to ABIS from a handheld device via mobile wireless communication vehicles, fixed towers, and a CENTCOM-operated server.[39] Table 3.4 summarizes the issues with the Transmit portion of the cycle.

---

[37] Interview with PM-Biometrics personnel, 2011. There is a cost of somewhat fewer matches because ABIS has three times as many stored files as the BAT system does.

[38] Records from SIPRNet are manually stepped down to NIPRNet so they can be fed into ABIS for matching.

[39] GAO, 2012. CENTCOM "was responsible for conducting the Last Tactical Mile pilot project during 2011 to provide the warfighter with a rapid response time on biometrics data submissions. In the pilot project, matching is initially against a biometrically enabled watch list stored on the warfighter's handheld device before searching against data stored in a stand-alone computer server in Afghanistan prior to transmission to ABIS—the authoritative database. The Last Tactical Mile project originated as a joint urgent operational need to utilize wireless infrastructure to transmit biometrics enrollments from a handheld biometrics collection device to a wireless communications tower ... Army officials told us that expanding the Last Tactical Mile pilot project across all of Afghanistan would cost approximately $300 million, in large part due to the number of wireless communications towers that would be necessary to provide connectivity ... DOD has not completed its evaluation of the Last Tactical Mile pilot project at the time of our review."

Table 3.4
Issues Associated with the Transmit Portion of the Biometrics Cycle

Cycle Issues	Implications	Risk	Reason
Field matches against ABIS subset replicated to each device.[a] Only SOCOM and BISA have satellite communication-enabled field access to BIMA for matching against ABIS records	Field matches against dated database Large expense to manually replicate data to BAT Inability to match against interagency data	M	HIIDE and BAT developed as QRCs without process to ensure interoperability with other systems and requiring adherence to standards Architecture will be modified with expected FY13 deployment of BAT-Av5 and will fully change JPIv2 fielded in FY16. No high capacity direct link between field & ABIS
Cross-domain solution only serves BAT portion of architecture	Uncoordinated activity as military services seek alternative to NGIC Marines tested SOCOM's web-based portal Inhibited ability to share with interagency partners	H	Non-BAT cross-domain solution not developed despite being warfighter priority

[a] CENTCOM comments, May 2012. CENTCOM states they submitted a JUONS to address the BAT data replication architecture that was never completed by PM-Biometrics.

Issues: Match

When we began the study, ABIS was reaching its throughput capacity for ingesting data and conducting match inquiries. An inability to handle matching requirements would either mean that biometrics are queued for processing—meaning the average biometric record has to wait longer (and what took minutes may take hours or days)—or given a lower priority to keep the rest of the inquiries flowing, with the hope that those biometrics will just be processed much later. However, the EA increased ABIS transaction capability from a peak of about 8,000 transactions per day to about 18,500 per day during our study, and plans call for further increases. Further, DoD's matching requirements are likely dropping: Mass enrollment is ending in Afghanistan and already ended in Iraq. Future large additions of data will likely arise from processing international partners' data, such as 800,000 fingerprint cards received from the Philippines, which will not be time-sensitive transactions. Nevertheless, the true impact of ABIS's limited throughput capacity is reflected in the inability to share robustly with interagency partners (see Share/Reference below).

Another factor lowering the efficiency of the matching process has been the prevalence of non-standard data, primarily from field devices (e.g., the HIIDE) that connect to the BAT. Non-standard data increase the work involved in the matching process and can lead to a failure to match when comparing individuals' fingerprints with latent ones collected from artifacts, such as recovered components of an IED. The EA has had time to revisit the QRCs in light of the longer-term future of the biometrics enterprise, but has not done so.

A third issue, which originated outside the EA's immediate control, is that matching algorithms for irises and faces are proprietary. In an era when, as BIMA officials note, irises are becoming more important as a matching tool among biometrics, the failure to develop a standard matching process that can be applied across the U.S. government may result in significant inefficiencies and potential errors in identifying individuals. This issue could be addressed by ensuring both DoD-wide licenses for algorithms and that all data created by devices meet DoD EBTS standards. Table 3.5 summarizes the issues with the Match portion of the cycle.

Issues: Store

A potential risk arises if the demand for record storage exceeds actual capacity. Although the capacity of ABIS was increased from approximately eight million to approximately 12 million records during our study, BIMA would have to face unpleasant options if plans to increase capacity in the future are delayed. One is that new biometrics will not enter the database or would be delayed indefinitely, e.g., cards from overseas partners being queued for entry. Another is that records currently in the database would have to be removed, at least temporarily. A third is that the information present in each biometric would have to be systematically truncated, leading to lower resolution matches (e.g., fewer fingerprint minutiae), or a reduced capacity for certain types of matches (e.g., against latent fingerprints). EA component staff members appear confident that sufficient funding will to be available for future ABIS storage capacity increases.[40]

A more serious issue is that the EA has made inadequate provisions for a COOP site, or "hot" backup. ABIS currently has a "warm" backup, which can come online in the event of expected downtime (e.g., a software upgrade), as long as the database managers have at least six hours' notice (which they may not necessarily have if the upgrade is forced upon managers as a result of a security breach). Plans are under way to establish a true COOP site following an initial requirement established in approximately 2007, but the go-live date continues to recede

Table 3.5
Issues Associated with the Match Portion of the Biometrics Cycle

Cycle Issues	Implications	Risk	Reason
Non-standard data sent for matches	Significant inefficiencies in the matching process and potential missed matching opportunities	M	Standards not enforced for QRC acquisition
Proprietary and incompatible matching algorithms for irises and faces	Significant inefficiencies in the matching process	M	Capability developed as QRC without process to address proprietary software issue

[40] ABIS stores all submissions as separate records regardless of whether there are multiple submissions for the same person. Questions were raised outside the EA about whether ABIS could be operated more efficiently if records were consolidated in a way that still preserved all data, but the EA believes efficiency would suffer.

following multiple proposals and at least one failed attempt. Table 3.6 summarizes the issues with the Store portion of the cycle.

Issues: Share/Reference

Although ABIS operates well with FBI databases, it has problems with its counterparts in DHS and within the IC. In the case of DHS, the problems include not only deficient throughput and storage capacity (at least for the time being), but also communications and protocol issues. The latter are being worked out through DoD-DHS interoperability meetings. Data that DHS collects are not directly matched against ABIS outside of regular exchanges of the BEWL and some pilot efforts of DHS components to transmit requests to ABIS. DHS's biometrics requirements have risen sharply with the introduction of the US-VISIT program wherein all foreign visitors entering on a visa are required to have their fingerprints checked against DHS's database, the Automated Biometric Identification System (known as "IDENT"). Consequently, the EA is under pressure to increase ABIS capacity specifically to meet DHS's needs, but the small intersection between fingerprints collected by DoD (other than those in the BEWL) and those collected by DHS has somewhat attenuated the need. The large price tag to increase ABIS's transaction capacity has also given EA components pause. Nevertheless, in its Functional Requirements Document, DHS said it needs to be able to send more than 100,000 biometrics inquiries per day to ABIS by early 2013.[41] DoD and DHS are working to address the interoperability problem between ABIS and IDENT and increase ABIS's capacity enough to allow two-way inquiries, but in the meantime, DHS only has access to a subset of DoD's data. This means unwanted individuals can conceivably slip through the cracks at U.S. ports of entry, if DHS sees them but does not know they have been previously identified by DoD (or by states that provide biometrics to the United States, such as the Philippines or the Dominican Republic).

Interoperation with classified biometrics data is a more difficult problem, since it cannot be solved simply by increasing throughput or capacity. ABIS is an unclassified system, fed by unclassified lines, and stored in unclassified spaces. Individuals who work on the final steps of the matching process are not always cleared to the highest level. Adding a cross-domain solution (also known as an automated "trusted guard") between the classified and unclassified

Table 3.6
Issues Associated with the Store Portion of the Biometrics Cycle

Cycle Issues	Implications	Risk	Reason
ABIS is reaching its storage capacity	Biometric data could fail to reach the authoritative database	M	Database upgrade failures and delays
ABIS has no "hot" backup A "warm" backup exists with a six-hour changeover	Possible loss of database access	H	Not prioritized by the EA, and lack of coordination

[41] BIMA comments, May 2012. BIMA observed that this 100,000 figure is at variance with the requirement of 14,000 in DHS's Systems Requirements Specification.

networks (as BIMA is doing to exploit SIPRNet connectivity) would not solve the problem. Thus, highly classified biometrics cannot be sent to or stored by ABIS and may not necessarily be examined by BIMA personnel.

Although both the classified databases and ABIS have access to the BEWL (which has only a subset of ABIS's data), neither has access to one another. This means records in the unclassified or classified databases could be incomplete and/or potentially fail to produce matches. Feeding ABIS's data into classified networks would provide a more complete picture for the IC, particularly with respect to people appearing in multiple geographic areas, which could ultimately inform strategic collection efforts. However, DoD has not taken any steps to replicate ABIS's data in the classified realm. The possibility of placing ABIS on the classified network exclusively was raised during our interviews, but this would inhibit ABIS sharing data with interagency partners because FBI's Integrated Automated Fingerprint Identification System (IAFIS) and DHS's IDENT both operate on unclassified networks. Table 3.7 summarizes the issues with the Share/Reference portion of the cycle.

Issues: Analyze and Decide/Act

Finally, there appears to be an organizational and conceptual mismatch between the biometrics activities and all-source intelligence activities. The EA, for instance, controls analyst billets even though BEI analysis is not a designated EA responsibility. The BEWL has also been managed by the EA even though it is not a designated EA responsibility, and there is little planning for the evolution of BEWL management, maintenance, and development. DIA is designated as responsible for the BEWL in DoDD 8521.01E but has only recently begun considering options for carrying out that function.

The seriousness of this problem is strongly related to what one thinks the role of the biometrics community should be. The biometrics enterprise could be simply a supplier of services to the IC (in the same sense that parcel delivery is a service to e-commerce), in which case its role is to convert disparate contacts into a person's unified history by indicating where ostensibly different individuals are really the same person under different aliases. The biometrics enterprise essentially

Table 3.7
Issues Associated with the Share/Reference Portion of the Biometrics Cycle

Cycle Issues	Implications	Risk	Reason
ABIS not interoperable with DHS's IDENT DoD and DHS are engaged in an ongoing interoperability effort to resolve this issue	Fewer matches, weaker analysis	H	Poor or lack of interagency coordination
ABIS cannot handle DHS processing load	Fewer matches, weaker analysis	H	Limited processing performance
Classified biometric files cannot be matched against ABIS records	Fewer matches, weaker analysis	H	ABIS cannot handle classified data; no provision has been made to mirror ABIS data into classified systems

Table 3.8
Issues Associated with the Analyze and Decide/Act Portions of the Biometrics Cycle

Cycle Issues	Implications	Risk	Reason
EA assignments do not encompass the full end-to-end biometric capability (EA controls preponderance of biometric analyst billets, even though BEI analysis is not a specifically designated EA responsibility)	COCOM BEI analytical needs may not be met Weakened analysis and underutilization of biometric data	M	Lack of requirements and coordination Lack of institutionalized BEI training
No plan for future BEWL management, maintenance, and development (BEWL has been managed by EA even though it is not a designated EA responsibility)	Decide/Act weakened	M	DIA has not assumed responsibility per DoDD 8521.01E

provides a bit of information, i.e., these two individuals are or are not the same person. If so, the biometrics and intelligence communities do not have to be tightly coupled.

Alternatively, one could envision the biometrics enterprise as an enabler of intelligence on people. If identity could only be established through physical contact (e.g., taking fingerprints, photographing irises at short range) then the amount of "new" information available on any individual is likely to be small. If, however, identity could be established at a distance, then the number of potential contacts would be much larger, and changes in intelligence collection such as automated surveillance would enable a richer set of information to be analyzed. In the latter case, integrated planning between the biometrics enterprise and the IC (or at least that part interested in unofficial individuals) would be compelling because changes in biometric technology could drive new or expanded strategic collections for the IC. The biometrics community should then understand the evolution of intelligence requirements, and the intelligence analysts should be aware of opportunities provided by advances in biometrics.

So the biometrics community is a service and nothing more, or it is an enabler and thereby needs to evolve concurrently with the IC. But the current posture that the biometrics community is an enabler that essentially evolves independently of the IC is inefficient and could degrade effectiveness over time. Further, the lack of integration of available biometric data into intelligence functions limits the ability of biometrics to inform decision-makers. Table 3.8 summarizes the issues with the Analyze and Decide/Act portions of the cycle.

CHAPTER FOUR

Status Report on the Biometrics Program

This chapter discusses the overall status of the DoD biometrics "enterprise." Biometrics activities in DoD are not an enterprise in the sense of a coherently planned, integrated effort. Instead, enterprise is shorthand for a set of disparate activities that grew from the bottom up. Our analytical approach to this section of the study was to consider gaps in policy, implementation, and related areas, and draw conclusions about the overall status across DoD. The greatest challenge in making statements about the status of the biometrics enterprise is that it lacks clear goals and measurable objectives against which we could judge its health and performance. Discussing the status is also difficult because of the recent and significant changes occurring, including the EA releasing the CDDs for BEC Increment 1 and JPIv2 for joint staffing, upgrades to ABIS, Executive Manager reorganization, the Navy continuing work on its Identity Dominance System, and the changing budget projections for FY13 and beyond.

The sections that follow describe our assessment of the overall status of the DoD biometrics enterprise and the specific gaps that informed our conclusions about the status. Similar to the assessment of the EA, the biometrics enterprise has worked in recent years to formalize activities and make biometrics more of an institutional capability. RAND expects that the required future status reports will track these continued changes and efforts to improve the enterprise's effectiveness.

Biometrics Capability Status: Red and Gray Forces

In general, DoD's biometrics capability for red and gray forces exists and has been supporting warfighter needs to some extent. The most frequent complaints were about inefficiencies in moving biometric data, lack of feedback to the field, lack of capabilities spanning the full range of missions and environments, and EA management issues. As noted above, there are multiple parallel architectures for collecting and transmitting biometric data to the authoritative database for matching. The original BAT scheme involved a lot of manual steps for processing, and the current BAT-HIIDE scheme is inhibited by a lack of communication capability and conformance with FBI fingerprint standards.[1] This has resulted in a lag in data transmission and has required a lot of manpower to keep the system running and to keep data as current as possible in ABIS and the regional servers. The parallel scheme used primarily by SOCOM also involves a manual step of moving data from SIPRNet to NIPRNet to perform the match inquiry in ABIS, followed by another manual step to send the match response. These systems have managed to provide for warfighter needs in the land theaters in Iraq and Afghanistan through hard work and sustained effort of every component in the cycle. However, the systems have not been modified in the last five years to incorporate greater automation and operate more efficiently, and the overall scheme for moving and handling data is largely unchanged. Significant changes are part of the BEC and JPIv2 PoRs, but they will not be fielded until approximately FY15 at the earliest. Further, an equivalent system for maritime missions does not exist, although an effort is in the initial prototype phase through the Navy's IDS, which was the basis for the future JPIv2 PoR.

A major reason for the current state of biometrics capabilities is that there is no common understanding across DoD about the purpose and boundaries of the enterprise. The governing documents are silent on the matter beyond acknowledging the simple existence of a capability and an authoritative database. In other words, DoD has no way to evaluate the sufficiency of the biometrics enterprise. A prior attempt to issue a DoDI was halted near the end of formal staffing around April 2010 because of language in the draft FY11 National Defense Authorization Act that would have significantly affected the biometrics community.[2]

The lack of boundaries led to a wide divergence of views within DoD about what is truly required for a biometrics program, with some people viewing it as a long-term, robust set of activities and others seeing it as only a niche capability. The divergence of views also resulted in biometrics capabilities being discussed and treated by some as ends unto themselves, rather than a means to the end of supporting missions. This is by no means unique within DoD, but an agreed-upon objective and boundary to the enterprise would enable better discussion of the issues raised throughout this report.

The lack of specific and measurable strategic goals and objectives for using biometrics has clearly hampered progress, particularly in the last five years as use has increased dramatically. For example, the PSA advocated for changes by the EA, but without concrete, agreed-upon objectives, the EA and the EA's components were essentially free to choose their own courses of action. BIMA, as the Executive Manager under the EA, produced a strategy and implementation plan, but it is not enforceable outside BIMA. The lack of progress is fairly clear in the years-long discussions about promulgating standards, enforcing standards, fielding interoperable equipment, and establishing a COOP site for ABIS, among others. BIMA noted the DoD BESP 2008–2015, issued August 27, 2008, as a source of goals and objectives. However, the strategic plan is vague and generally tracks DoD's broad mission of supporting the warfighter.

---

1 The expected upgrade from the HIIDE to the SEEK will alleviate the fingerprint issue, but it remains to be seen how communication will be handled.

2 Comments from OSD to RAND on the draft of this report, May 2012.

Also, the strategic plan was issued in 2008 while BIMA was still known as BTF, U.S. troops were still present in Iraq, and no formal PoRs had been developed.

The inability to make progress is also hampered by the weak governance structure. The PSA as EXCOM chair has no formal directive authority, and the same topics arise repeatedly in working group and EXCOM meetings without explanations for lack of progress or a consensus on deadlines. The ultimate consequence is that the biometrics capabilities are slow to respond to changed circumstances, advances in technology, and increased demands for information both inside and outside DoD. In some instances—most notably the Navy's IDS—the military services and COCOMs have initiated their own biometrics-related efforts in the face of this slow progress. The military services often try to preserve freedom of action in all aspects of DoD governance, with the effect being lack of consensus, and biometrics may be no different.

Consequences for Changing Environment

The structural inefficiencies and lack of goals have had a direct effect on the current status. DoD's biometrics capabilities have not transitioned from "quick reaction" systems to a coherent set of programs and activities, despite apparent support for long-term use. The fielded systems are formally designated as QRCs, but a system that has existed in some form for ten years—with the current iteration essentially unchanged for over five years (except for ABIS, which is just over three years old)—could potentially be better organized, or at least on a path to standardized implementation.

Three main factors appear to have contributed to the continuation of QRCs: OCO funding, focus on devices, and opaque prioritization of JUONS responses. As noted above, the enterprise has been largely funded by OCO supplemental appropriations and is not part of a base budget. This prevents long-term planning for upgrades, new development, and potentially broader applications. Biometrics activities have also been focused on procuring devices rather than achieving measurable outcomes, including long-running conflicts over BAT upgrades and handheld devices. The need to address urgent operational needs at the tactical level, including negotiating over technical aspects of devices, diverted attention from the more important question of the value provided by biometrics capabilities. Finally, the process used by the EA to prioritize responses to JUONS was opaque, drawing complaints from the COCOMs that the most important needs were not being addressed. As noted by Headquarters, Department of the Army personnel, JUONS were received and disseminated to the appropriate organization along with OCO funding to support, but no system existed for several years to account for how the money was spent in attempting to fulfill the JUONS.[3] According to these same personnel, the Army now has a process to track JUONS, but the big push to field biometrics systems for wartime is largely over.

---

[3] Interview with Headquarters, Department of the Army personnel, 2011.

Many of the disputes about maintaining and transitioning the QRCs will be resolved de facto by the decline of OCO funding and expected DoD-wide budget cuts. This means the Army, in particular, will have to make hard choices when looking for trade-offs within its base budget to support the new biometrics PoRs. This also means new disputes are likely to arise within and among the military services and COCOMs over the scope and scale of biometrics capabilities going forward. DoD's shift away from Iraq and Afghanistan means a reduced need for biometric matching that supports tactical operations for conventional forces, but other biometrics requirements appear to be expanding. Specifically, deficiencies in DoD's current biometrics enterprise limit the ability to share information with and support interagency partners.

Biometrics Capability Status: Blue and Gray Forces

While most of the biometrics resources have been applied toward red-force missions, DoD has made some effort in biometrics activities for blue (friendly) and gray (third-party national) forces, but this is still in very preliminary stages, with many unresolved questions. Under DoDD 8521.01E, the OUSD(P&R) is responsible for maintaining "the authoritative source and reference archive for biometric information collected from friendly individuals issued a DoD credential."[4] Holders of a CAC must provide a fingerprint to obtain and renew their cards, but DoD does not use the fingerprint as a verification measure for facility or computer network (i.e., logical) access. Further, DoD did use the EA's systems for gray biometrics in Iraq and Afghanistan, but it remains to be seen whether this application is institutionalized and adapted to missions outside of large conventional force operations dealing with large foreign civilian populations.

Most important, shrinking budgets will require a clear justification for any new investments. Biometrics are well understood as strong verification measures, but their associated costs and benefits are not understood. For example, DoD has not assessed the risk associated with not using biometrics for blue-force facility and logical access. In other words, what percentage of actual or potential penetrations of DoD facilities and systems could be averted by using biometric verification? Further, the cost of implementing such measures has not been identified, with only preliminary pilot studies at DoD facilities under way or under consideration during our research. There are also no indications of what trade-offs would be made to create space in a DoD component's base budget for new blue-force biometrics activities. Like most technology, biometric equipment costs will decrease to potentially allow local operation and maintenance funding (e.g., local installation budgets) to implement biometrics. However, the risk is that data, and possibly non-standard data, could end up in disparate databases that are not interoperable, unless the PSA and EA maintain visibility into the various implementation efforts.

---

[4] DoD, 2008a paragraph 5.4.4.

The practical effect of blue-force biometrics measures is also a significant and yet-to-be determined consideration of cost. Any slowing at the gates of military bases presents risk, and slower entry inside buildings and for computer access—caused, for example, by false negative responses—are potential costs in terms of productivity. Some percentage of DoD and non-DoD personnel will also object to the use of biometrics for these purposes, creating potentially lengthy internal policy debates.[5]

Also, advocating for large-scale implementation of blue-force measures is more difficult given the current use and apparent success of DoD's two-factor authentication for logical access with the CAC and a personal identification number (PIN). In interviews, personnel in the DoD Office of the Chief Information Officer (OCIO) cited a 40 percent drop in known NIPRNet compromises because of the CAC-PIN authentication, and the OSD-CAPE described blue-force biometrics as "a solution looking for a problem."[6]

The strongest argument in favor of new investment appears to be for vetting and physical access for gray-force personnel at overseas locations, e.g., host-country nationals who work on DoD facilities.[7] However, a biometrically focused risk assessment and cost-benefit analysis has not been performed (or was not made known to RAND) that would make a clear case for new capabilities.

Identified Gaps Supporting Status Report

To reach the status determination, we combined our analysis of the biometrics cycle (see Chapter 3) with an analysis of gaps. Specifically, we identified gaps in DoD policy, requirements, implementation, and analysis, that we assess as presenting a high risk ("H") to the ability of the biometrics enterprise to meet DoD needs currently or in the future; or medium risk ("M") of affecting the ability of the enterprise to meet DoD needs in the future. Like the discussion of the biometrics cycle, these are subjective differentiations, but we believe the data support our primary determination that these gaps pose risks to the enterprise.

We identified gaps across the spectrum of high-level functions, with several gaps likely interconnected. We believe the first four gaps are the most important, and effort on those gaps will have a significant impact on the overall direction of the enterprise.

Strategic Goals (H)

The most important gap is DoD's strategic goals for biometrics. As noted above, a vision statement exists, but it is vague and lacks clear, measurable objectives that could be used as a basis for long-term planning. Setting the strategic goals will clearly help some of the other gaps dis-

---

5 True legal limits are not an issue, but views of DoD's Office of General Counsel and the policy community would still be a factor.

6 Interview with OSD-CAPE personnel, 2011.

7 Host nation privacy laws would also be a potential factor for this use.

cussed in this section and will help some of the longstanding conflicts that arise in the working group and EXCOM meetings. In other words, once the boundaries and direction for the enterprise are established, chances for success are improved because success will finally be defined.

Cost-Benefit, Risk, or Other Analyses (H)

The lack of cost-benefit, risk, or other analyses supporting biometrics investments is another serious gap. As far as we could determine, no objective measurement or understanding of the real or potential value of DoD biometrics exists.[8] The lack of such analyses makes it nearly impossible to determine a return on investment—or, more accurately, current results provided by resources expended—beyond anecdotes about "bad guys" identified, e.g., the "Hit of the Week" and "High Value Individuals" encountered. Everyone understands that biometrics activities are providing some value, but it has not been quantified and may not be possible to quantify. Most important in light of anticipated budget cuts are results from outlays for biometrics activities as compared to other investment opportunities, i.e., trade-offs. The CDDs for BEC and JPIv2 released by the Army in February 2012 contain cost-benefit analyses, but these are analyses of the costs and benefits of different acquisition options for biometrics tools, not an evaluation of costs and benefits of acquiring or not acquiring biometrics tools.

Implementing Instructions, Doctrine, and Strategy-to-Task Implementation Plan (H)

Partly as a result of not having overall strategic goals, the DoD biometrics enterprise lacks implementing instructions, doctrine, and strategy-to-task implementation plans, a shortcoming that is already hampering the enterprise. DoD has a well-established progression of key documents, including DoDDs, DoDIs, and Planning Guidance, that direct and guide the DoD components in executing missions. This particular gap is known and understood within OSD, as a draft DoDI is in the works at the time of writing this report that coincides with potential revisions to DoDD 8521.01E, along with the Doctrine, Organization, Training, Materiel, Leadership and Education, Personnel, and Facilities Change Recommendation (DCR) for BEI noted below. These pieces of policy are critical for both current enterprise effectiveness and future planning, especially in light of the new DoD PoRs BEC and JPIv2 and the Navy PoR IDS.

DoD Intelligence Strategy (M)

We believe the IC is likely to be the biggest consumer of biometric data in the future, and the lack of a DoD intelligence strategy for biometrics could affect the overall performance of the enterprise. Biometrics is correctly viewed as contributing to all-source intelligence, but thus far coordination among military services has been idiosyncratic. NGIC has been shouldering

---

[8] BIMA comments, April 2012. BIMA cited their ongoing identity authentication study due in September 2012 that is "focused on data from a technical literature survey, scenario and operational pilots to analyze the justification for directing resources towards biometrics as an identity/identity-enabler for physical access control." Results from such a study may be valuable for DoD. BIMA also cited an International Organization for Standardization (ISO) framework, which according to ISO, "ISO/IEC 19795-5:2011 specifies a framework for testing and a grading scheme for reporting the performance of a biometric system suitable for use in access control applications." Again, this is a potentially useful reference for DoD.

the BEI analytical burden for Iraq and Afghanistan, but conflicting messages about whether the Army will perform any non-Army analysis in the future—and what share DIA may perform—keeps the issue unresolved. The lack of a DoD intelligence strategy, and how it would inform requirements, may leave an important part of the biometrics picture muddled. A DCR is under consideration at the time of this report to institutionalize BEI. This DCR could help with the implementation of an intelligence strategy and with the broader gap of instructions, doctrine, and plans discussed next.

Requirements (M)

The current state of requirements for biometrics could have a negative impact on the enterprise over the next year or two. The CDDs for BEC and JPIv2 were released by the Army in February 2012 to begin joint staffing. The programs cannot begin to provide any capability until they reach Milestone B in the acquisition process. The Army estimates the PoRs will reach Milestone B in the formal acquisition process in late FY13, which is in keeping with the commonly held estimate of about one year after CDDs are released. This means the Army cannot fully add them to their base POM until the FY14 budget beyond an existing budget "wedge" for research and development, i.e., a budget placeholder used before the true size and nature of a program is known. The result is that the existing, legacy QRCs of ABIS and the BAT-HIIDE systems (soon to be BAT-SEEK) will be the only available tools in the interim, and they do not have resources budgeted beyond FY12 because of the reliance on OCO funding. Further, the Marine Corps said it is in danger of losing FY13 through FY15 procurement funding because the lack of the JPIv2 program leaves it without a program to purchase against.

Requirements Coordination (M)

Requirements coordination has apparently improved in recent years, but it may still cause problems for the biometrics enterprise in the future. As noted above, both TCM-B&F and BIMA conducted requirements-related activities in separate working group-type forums. While we heard conflicting accounts of the exact nature of discussions in those forums, the fact is that sources interviewed inside and outside the Army were unsure of who was responsible for requirements coordination. Further, several interviewees voiced complaints that non-Army requirements were either not solicited by the EA components or that non-Army requirements that were brought forward were subsequently not included in requirements documentation. The substance and scope of these claims are unclear, but the fact that personnel voiced them confirms a gap in requirements coordination. TCM-B&F stated that it attempted to collect requirements from across DoD and incorporate them into the CDDs for BEC and JPIv2, so the true test of this assertion will be the forthcoming joint staffing process. Further clarifying roles and responsibilities in the EA structure would also help to facilitate future coordination.

Programs of Record (M)

The PoRs for BEC and JPIv2 are a frequent point of discussion in the working group and EXCOM meetings and were raised often during interviews. The PoRs remain a gap that could affect the enterprise in the future, especially if Milestone B is reached after the Army's estimate of late FY13. The CDDs for the PoRs required approximately three years for completion and submission to the Joint Staff in February 2012. We heard conflicting claims about the reasons for the long process, including changing requirements, additional studies, and bureaucratic slow-rolling. Regardless of how, or if, any of those claims contributed to the delays in completing the CDDs, the PoRs are a significant gap that may negatively affect the enterprise if joint staffing is not completed by the end of FY12 as the EA estimates. The G-8's proposed stopgap measure of continuing to rely on the currently fielded QRCs is still in question because they were not funded beyond FY12 (as of early February 2012). Without FY13 funds, ABIS would not operate, and personnel in Afghanistan and other locations, as well as interagency partners, would not be able to send inquiries to DoD's authoritative database for red and gray forces.

Further, there is a gap between the nature of the PoRs as described in the CDDs and expectations for future biometrics needs. As discussed above, the CDDs are geared toward formalizing and expanding the types of activities in Iraq and Afghanistan, such as mass collection of biometric data from a civilian population in a permissive environment. However, all indications right now are that biometric use will instead be focused on intelligence, counterterrorism, other threat networks, interagency law enforcement support, and international partner cooperation. Thus, the PoRs may be delayed further because they have to be significantly modified to meet expected needs.

Data Standards (M)

Data standards are another subject that was discussed frequently in meetings and interviews and constitutes a gap that may limit future biometrics enterprise effectiveness. BIMA (and its predecessor organizations) spent significant time examining, advocating for, and issuing documentation on standards for biometric data. However, BIMA lacked any authority to enforce the standards and has had to accept data in all forms from all DoD sources.[9] Further, interviewees asserted BIMA (or a predecessor) made previous abrupt, unannounced changes in standards that affected ongoing technology procurement/development. The lack of clarity and specificity about standards led to some of the interoperability problems that are a regular point of discussion. This issue appears to be resolved as to the future PoRs that would have enforceable standards, but those PoRs are not likely to be fielded before FY15. PM-Biometrics is developing BAT-Av5, which is expected to adhere to EBTS standards, but the extent of this improvement is yet to be determined because PM-Biometrics was required to deploy the new SEEK II handheld devices compatible with the current non-EBTS compliant version of BAT.[10] In the interim, given the lack of direction from the EA regarding standards implementation, the PSA is directing compliance with the standards, and at least one implementation plan the PSA expects to take effect at some point in the future will help work toward improved interop-

---

[9] BIMA comments, April 2012. BIMA stated that it "enforced the standard for those who chose to follow it." We note that the ability to choose whether to follow a standard is the definition of unenforceability.

[10] PM-Biometrics comments, April 2012.

erability. Any shift will have to be carefully implemented to ensure no loss of data or currently fielded capability.

Architecture Policy (M)

The lack of an enforced architecture policy has led to the parallel biometrics cycles for conventional and other forces, and this gap could complicate future attempts to change the enterprise. The primary architecture involving the BAT is essentially unchanged in the last five years, with the biggest change being the addition of a HIIDE that can download data to the BAT via a physical link. This architecture was initially manpower-intensive because it required manual replication to servers. According to PM-Biometrics, data replication has since become automated, but the scheme still has lags in data being added to ABIS because the HIIDE must be physically attached to a BAT device to transmit data. One example of this problem was an individual who was encountered and enrolled three or four times before his biometric data were matched to previously unidentified fingerprints from multiple IEDs. In the case of the HIIDE's origin, PM-Biometrics stated it was fielded "as a separate QRC solution by PM Army Space Programs Office ... in 2005, not as part of the BAT QRC. PM DoD Biometrics did not take responsibility for HIIDE until FY10" when it was merged with the BAT QRC.[11] The need for rapid fielding was used to explain the lack of a strategic or systematic approach to architecture and technology development. However, biometric technology for red- and gray-force applications has been fielded for ten years, DoD made a commitment in principle to maintaining the capability in the future, new PoRs are in the works, and PM-Biometrics is working on a BAT upgrade that would address some architecture issues. Architecture should be addressed systematically to eliminate lag and parallel data transmission, and increase efficiency.

Tactical Communication (M)

Tactical communication is a gap that may affect the enterprise in the future but should be addressed by the appropriate part of DoD with input from the biometrics community. DoD focuses a great deal of effort on improving battlefield and theater communications, especially with limited bandwidth and significant growth in the types of data being collected and transmitted in the field. The biometrics community has spent considerable time and effort on communications, even though it would be in the purview of other DoD organizations.[12] As noted above, the EA helped fund the LTM pilot effort to improve transmission of biometric data.

OCO funding was the enabler for biometrics community work on communications, and, as has been mentioned, that funding source is disappearing. The biometrics community should remain engaged with the communications community to ensure requirements and shortfalls are addressed, but it should limit work on tactical communications to maintaining a capability that can interface with available DoD systems.

---

[11] PM-Biometrics comments, April 2012.

[12] PM-Biometrics comments, April 2012. PM-Biometrics stated that it had not "spent time, effort, and funds on biometric-unique" communications.

Training and Education (H)

Training and education for biometrics are a significant gap that is currently diminishing the effectiveness of the enterprise. Pre-deployment training with biometrics tools has been ad hoc, despite a decade of red-and-gray-force biometrics collection and use, and, as discussed above, no formal training and education effort will happen without a formal PoR. The lack of training has contributed to reduced effectiveness and efficiency in the biometrics cycle, particularly at the point of collection. For example, we were informed of several instances of fingerprints being attributed to the wrong hand. EA personnel also described inconsistent data entry in the field for records of enrolled individuals, which hampers analysis and decisions about what to do when a match is found in the database (often referred to as the "so what" component of a biometric match). In addition to training in the use of biometrics technology, DoD may need to educate senior officers about the use and value if the intent is to keep biometrics as part of the full set of missions.[13]

TCM-B&F has attempted to develop some virtual training modules outside the scope of its responsibilities, but the true resolution of this gap will only come from the proper component of TRADOC with formal training and education. TRADOC could produce a common program of instruction and provide it to the other military services.

U.S. Government Strategic Goals (H)

Like DoD, the broader U.S. government lacks clear strategic goals for using biometrics. This hampers the current DoD enterprise, development of DoD strategic goals, and planning for future activities. As noted above, DoD's focus has already shifted away from Iraq and is shifting away from Afghanistan, the two places where biometrics collections have been the greatest. DoD will be expected to continue contributing—likely at increased levels—to interagency missions, strategic collections for intelligence, special operations, and partner nation activities. However, DoD is limited in its ability to determine the level of effort needed and to plan and program for future resources without clear U.S. government goals. The lack of goals also contributes to continuing issues of interoperability and information-sharing protocols. As noted above, DoD and DHS do not have the ability to directly share data, and DoD data lacks sufficient information to easily separate information that is restricted from that which can be shared with DOJ and DHS. DoD must continue to improve coordination and work in interagency forums to develop these goals and determine the scope and scale of DoD's efforts to support non-DoD activities.

Intelligence Community Strategy (M)

Similar to the U.S. government's goals gap, the IC lacks a formal biometric prioritization and collection plan, which affects DoD's biometrics enterprise. We believe future biometric

---

[13] GAO, 2012. According to GAO's most recent report on biometrics, "existing biometrics training for leaders does not instruct unit commanders and other military leaders on (1) the effective use of biometrics, (2) selecting the appropriate personnel for biometrics collection training, and (3) tracking personnel who have been trained in biometrics collection to effectively staff biometrics operations."

data collection will be geared more toward intelligence use, and DoD has begun considering input for a future IC plan. The IC needs to continue developing its goals for biometrics use, particularly for strategic collections, to ensure maximum cooperation among intelligence components. This is especially important in the area of biometrics because of the potential for so many different organizations and agencies to be collecting biometric data (as compared to something like signals intelligence [SIGINT] that has a clear lead agency).

Implementation of Interagency Agreements (M)

DoD has engaged with DOJ and DHS to facilitate biometric data-sharing through interagency agreements, despite the absence of explicit U.S. government goals, but these agreements have not been fully implemented. DoD and DOJ directly share biometric data between ABIS and IAFIS, which was likely facilitated by the original ABIS being based on the structure of IAFIS and the ABIS hardware residing in the same facility as IAFIS. Yet, the lack of a direct connection to DHS's IDENT system means DoD does limited, indirect sharing of data with that department.[14] BIMA highlighted the fact that it had sent the entire BEWL to DHS, which had "requested additional data from DoD ABIS for [approximately 320,000] files of which they estimated would take at least 90–120 days to ingest into IDENT." DoD and DHS are engaged to improve interoperability and ultimately create a direct ABIS-IDENT connection. BIMA stated the matter "is mired in DHS policy" and claimed they have been waiting for 10,000 test files from DHS for "more than a year."[15] DHS agreed the test files have required more time than expected to secure the necessary permission and ensure privacy protection, but they said the time that has passed is closer to six months.[16] Multiple interviewees noted the lack of clarity regarding exactly how many inquiries DHS hopes to send to ABIS, and interviewees also highlighted the issue of how to pay for expanding ABIS's capacity to accommodate DHS and DOJ inquiries. Resolving the scope and scale of interagency sharing and the extent to which the process will be automated will be critical for DoD.

---

[14] Some DHS personnel have pilot-tested equipment on loan from DoD that allows a direct link to ABIS. However, as stated by DHS personnel, this is a very small number of inquiries.

[15] BIMA comments, April 2012.

[16] Comments from DHS to RAND May 2012.

CHAPTER FIVE

Conclusions, Options, and Recommendations

Conclusions

DoD's biometrics systems function, but DoD lacks an agreed-upon set of goals for biometric use and a coherent end-to-end biometrics capability. In part, the EA had to respond to urgent operational needs with QRCs for Iraq and Afghanistan. Our review of existing capabilities and the statements of personnel interviewed led us to believe that OCO funding dulled the incentive and accountability that normally leads to efficient resource management and the development of an integrated system. However, the biometrics enterprise lacked strategic focus, which prevented it from laying a path to a coherent, integrated, and cost-effective system. Most of the progress toward a more systematic approach has come in the last year as the process has finally begun in earnest to create formal PoRs for an institutional capability. Before now, most of the activities were overly focused on procuring devices without consideration of an end-to-end capability and the value being provided. This led to multiple devices being fielded that were not completely interoperable and did not provide data at the standard level of quality.

Despite technical incompatibilities, inefficiencies, and interpersonal conflict at all levels, personnel in theaters, West Virginia, and elsewhere, persevered to meet operational and tactical needs with the systems fielded. Their hard work demonstrated some value of biometric technology, helped identify IED makers, and helped manage the large populations of detainees, among other functions.

The way forward for DoD biometrics is unclear. DoD's biometrics activities grew and were executed from the bottom up primarily to support a large conventional force engaged in mass collection of people's information. For the future, budget constraints and the new DoD Strategic Guidance mandates clear understandings of the returns on biometrics investments, risk management, and potential trade-offs to create budget space for biometric activities.[1] Key to continued funding for biometrics will be making the case that biometrics provide the most cost-effective approach for addressing designated missions and threats, based on a firm analytical foundation beyond anecdotes of success from the field. This will be especially important because the biometrics community is still working to establish and define itself. Nevertheless, the opportunity is available to make the case for maintaining and possibly expanding the bio-

[1] DoD, *Sustaining U.S. Global Leadership: Priorities for 21st Century Defense*, January 2012a.

metrics enterprise with DoD's apparent shift to a non-wartime focus on intelligence, counterterrorism, counterdrug, interagency law enforcement support, and international partner support. Biometrics may be an enabler for a smaller military engaged in a broader range of efforts worldwide.

The EA is looking to fund the DoD biometrics system out of the Army's base budget for FY13 and beyond, and senior Army leadership will likely increase the pressure for greater acquisition and operational efficiency. There are signs that cost saving could come at the expense of potential improvements in capability and support for non-Army users. For example, the Army's January 2012 projections showed no funding for ABIS beyond FY12, and the Army has previously issued official communication indicating that it will stop providing analytical support for non-Army DoD components for BEI. The Army's cost-reduction efforts risk exacerbating current conflict among military services and could lead to proliferation of service-specific programs that are developed independently and are insufficiently interoperable. Throughout this study, multiple Army personnel agreed that some of the issues highlighted in the assessment and status report sections above have not been addressed.

DoD must make significant changes to the strategic goals, organizational structure, technology acquisition, and resources to utilize biometric information. The current primary use of biometrics to support a large conventional force is shrinking, and DoD appears ill prepared to address future challenges and support growing interagency missions. The starkest example may be the fact that the PoR documents for BEC and JPIv2 are geared toward continuing support to large conventional operations that will be over by the time the programs' technologies are fielded in FY15 or later. DoD must decide what is truly needed for, and expected of, an end-to-end biometrics capability. Even as these goals are being established, the EA should ensure its organization does not have overlapping and conflicting responsibilities within the Army. Under the current EA structure, the disparate components working on biometrics capabilities only meet organizationally at the EA himself (SECARMY), which reflects a "stovepiped" organization.[2] One valuable change would be to make the Executive Manager a true manager of all EA functions, not just the portion that is currently BIMA, so that one person (other than SECARMY) is ultimately accountable for the full scope of the biometrics program. BIMA voiced its support for this notion when commenting on the draft of this report by advocating to "empower" the new Executive Manager in the Office of the Provost Marshal General (OPMG).[3] BIMA also believes more care needs to be taken to document changes in procedures, and the Executive Manager may be the logical person to be responsible for documenting and reporting to the PSA all significant changes throughout the EA components.

Changes within the Army, particularly for technology acquisition and the analysis and use of biometric data, must incorporate the fact that the new BEC and JPIv2 systems are

---

[2] The EA's organization mostly represents the standard within DoD of separating requirements, system development, and user functions, and a great deal has been written in other contexts about "bridging" among these stovepipes. However, the added role of BIMA and the unclear or overlapping responsibilities exacerbated the usual stovepipe issue.

[3] BIMA comments, April 2012.

unlikely to be fielded before FY16, and there is a chance both PoRs will ultimately become unfunded requirements because of budget constraint. This means currently fielded systems will need to be maintained and upgraded for some time, without the structure of the formal acquisition cycle. In other words, any changes within the EA should focus on managing cost effectively, prioritizing resources to identified needs, and increasing process efficiency.

A common theme throughout this study was the fact that biometrics have been treated as an end unto themselves rather than the means to the end that they are. It also appears, at least within OSD, that there are artificial divisions between biometrics, forensics, and other red-force-related identity operations. For example, much of the use of biometrics in Iraq and Afghanistan was a forensic function of trying to match latent prints (e.g., from IEDs) to people encountered by U.S. forces in the field. Thus, OSD should consider placing all policy functions for red-force identity operations under a single PSA.

As part of a broader examination of possible changes to the biometrics enterprise, OSD should also consider whether another DoD component is better suited to assume the role of EA given the expected reduction in conventional, ground-force collection of biometric data after Afghanistan operations end. Some options for the EA are presented below. Our interviews also demonstrated a gap between user-perceived needs for a working end-to-end biometrics cycle and the EA's approach (justified by language in DoDD 8521.01E) of being strictly a technology provider. This gap leaves no organization accountable for ensuring the biometrics cycle is functioning to meet user requirements.

Finally, DoD and the EA lack outcome metrics for assessing the value, effectiveness, and efficiency of biometrics activities. Metrics are regularly cited as critical for DoD's wide range of activities, and the field of biometrics is no different. The EA has created some output metrics to try to measure the efficiency of activities, most notably ABIS data collected by BIMA. However, metrics useful for external assessment that demonstrate outcomes and value have not been developed and used to this point. The changing scope of use for biometrics and the need to justify continued investment as part of base budgets demand better metrics, particularly for entities assessing and evaluating the biometrics community. Appendix C provides an extended discussion about metrics and the types of metrics that could be applied to the biometrics enterprise.

Options and Recommendations

Based on what we see as unresolved issues in the enterprise and the apparent future needs, we developed a set of options related to the EA and DoD writ large, and we provide our recommendations to improve biometric activities and develop metrics to evaluate the biometrics cycle.

Options for EA

This report has argued that the management of the biometrics enterprise by the EA has revealed apparent shortcomings. As ground operations in Afghanistan wind down and biometrics has to be funded out of military service budgets rather than OCO accounts, the environment will change, and not necessarily in ways that favor leaving EA management where it is. First, the pressure on Army funding creates incentives to shortchange the requirements of other military services and defense agencies in favor of those of the regular Army. Second, SOCOM collections and increased sharing with foreign partners (e.g., the Philippines) will overtake conventional ground force collections in size and reduce the Army's direct role in biometrics.

Together, the changes justify revisiting how the EA is managing biometrics responsibilities and potentially the decision to vest the EA responsibility with the Army. Accordingly, we consider five options: status quo, reduced Army EA responsibilities, OSD as EA, joint program, and SOCOM as EA.

Status Quo: The primary advantage of keeping SECARMY as the biometrics EA is that any move to change the locus of the EA incurs transition costs and could mean losing institutional knowledge. A battle over EA placement might absorb institutional energy better spent on addressing specific biometrics issues. Further, the EA has acknowledged some shortcomings and is taking steps to improve efficiency and effectiveness, such as creating formal PoRs and reorganizing the Executive Manager under the OPMG.

The risk of retaining the status quo is that any problems in the biometric enterprise that have roots in how the organization is structured would be difficult to overcome. For example, ongoing issues among EA components, including the conflict between BIMA and PM-Biometrics over the ABIS upgrade failures, may continue if the perceived interests of the organizations do not change. Any institutional problems that have slowed progress on addressing fundamental problems within the biometrics enterprise would require directed management attention.

After we completed our research, we were notified that the EA is moving BIMA's Executive Manager and ABIS responsibilities under OPMG. This move brings together biometrics management with criminal investigations (many of which rely on evidence based on biometrics). The shift in management may provide some efficiency in terms of staff, but it would not necessarily alleviate the issue of the Executive Manager having no authority over the other EA components—nor does it necessarily reduce the potential for further conflict with PM-Biometrics, nor address concerns about meeting user needs.

Reduced Army EA Responsibilities: This option calls for reducing the responsibilities of the EA to managing the database and procuring devices, while clearly placing the policy responsibilities, planning, and interagency/international liaison functions in OSD.

Such a move would have the benefit of centralizing the policy direction for biometrics, thereby allowing DoD to work more easily with external partners. It also makes clear that decisions are made on a DoD-wide basis. Finally, it addresses shortfalls that we observed in the current EA's ability to plan for the evolution of the biometrics program.

The downside may be that it would create the possibility of disconnects between policy (OSD) and implementation (the Army), although disconnects appear to exist within the Army itself.

OSD as EA: The benefits and costs of shifting the EA responsibility to OSD are the converse of keeping such responsibilities with the Army. On the plus side, it may better address the interests of all stakeholders. On the minus side, such a shift will incur transition costs and could lead to a political fight. Further, it would saddle OSD with direct responsibilities for systems management and acquisition, which would cut against DoD plans to freeze the size of OSD and reduce costs as part of broader efficiency goals. It would also require OSD to budget for such a capability, and would, in all likelihood, require the creation of a separate agency (since OSD's manpower slots are limited). Implementing this approach would not necessarily require a legislative change as long as SECARMY clearly remained EA for biometrics IA. However, creating a somewhat artificial split by making someone other than SECARMY the EA for biometrics (minus IA) could be problematic.

Joint Program with SECARMY as EA: The benefits and costs of this option are similar to those of establishing OSD (or a defense agency) as an EA. On the plus side, it represents an even-handed all-stakeholder approach that has existing precedent within DoD. On the minus side, it is more costly to set up a separate program for what is a relatively small enterprise, which would eliminate the potential benefits of being part of a large organization.

SOCOM as EA: Moving responsibility to SOCOM would put it in the hands of an organization that has a keen interest in biometrics, has a direct understanding of user needs and experiences, and spent its own money to get quick results from its own biometric submissions. To the extent that SOCOM includes members from all military services, it may also be considered joint. Conversely, SOCOM (like the Army) is just one customer—which creates the risk that SOCOM personnel would favor their own operations at the expense of others. SOCOM has a highly specialized mission that may be too narrowly focused to properly consider DoD-wide biometric needs. Indeed, if the distance between SOCOM's needs and those of the military services is greater than that between the regular Army's needs and everyone else's, then the effects of taking a parochial approach could be worse. Finally, such a switch may be politically difficult, especially to the extent that SOCOM's rising status vis-à-vis conventional forces has created tension with the military services. Like OSD as EA, no legislative change is needed, but the same issue remains of a potentially artificial division with SECARMY as EA for biometrics IA.

We chose not to make a specific recommendation because of the uncertainty with making a major EA change.

Options for Information-Sharing

DoD is likely to put increased focus on interagency and international information-sharing as it shifts away from large, conventional OCOs. Consequently, the biometrics enterprise must be prepared to handle a larger number of transactions and inquiries with DOJ, DHS, and the IC. As noted above, ABIS already has a direct link to DOJ's IAFIS, and the greatest projected growth appears to be in DHS's databases. This is further highlighted by the ongoing effort to improve DoD-DHS biometrics database interoperability beyond DHS receiving the BEWL from DoD on a regular basis.

In the near-term, DoD should look for ways to increase the amount of data shared with DHS by including non-BEWL data. In the near- to mid-term, DoD should figure out how to add transaction capacity to ABIS and establish a direct link to DHS that will support the anticipated number of inquiries. The expected deployment of BEC in FY15 or later should incorporate continued potential growth in data-sharing, but the uncertain future funding for BEC means the EA should begin planning now to keep ABIS viable with an increased sharing capacity.

The IC is also likely to increase demand for biometric data, and DoD needs to add both capability and capacity to share information. The EA has been aware for some time of a need for an automated cross-domain transfer capability to move data between ABIS and classified systems, and work is ongoing to implement an initial solution in FY12. DoD should consider this a high priority along with expanding capacity for data-sharing in the near term. In the future, DoD should consider adding a classified store-and-match capability that contains all of the unclassified data in ABIS (or successor systems) and is maintained and operated by the EA. The simple presence of certain biometric data on a U.S. system could potentially reveal sources and methods, so these data need to remain protected. However, DoD should avoid having separate organizations operating the classified and unclassified databases to ensure full interoperability.

Options for EA Responsibilities in ABIS/BEC Development and Operation

With interagency information-sharing taking a more prominent role, the functionality of the authoritative database (ABIS/BEC) becomes more important, and DoD should consider how best to manage development and operation. One option is to maintain the status quo of the split between PM-Biometrics developing, upgrading, and maintaining ABIS, and BIMA (West Virginia) operating ABIS and doing some troubleshooting. To improve efficiency and effectiveness, the EA needs, at a minimum, to force increased cooperation between PM-Biometrics and BIMA and to implement genuine accountability measures for future failures. This approach avoids telling the EA how to manage its responsibilities. However, the history of conflict between the two organizations does not bode well for future cooperation as evidenced by the multiple failed attempts to upgrade ABIS. The division between PM-Biometrics and BIMA will likely lead to continued finger-pointing and difficulties with the eventual fielding of BEC. As noted above, the EA is planning to move BIMA's responsibilities under the OPMG in FY12, but this would not necessarily alleviate the conflict with PM-Biometrics.

Another option is to co-locate the ABIS/BEC development and the operations and maintenance (O&M) personnel, which would likely mean moving more PM-Biometrics personnel to West Virginia. This would at least potentially facilitate improved cooperation and prevent claims from both groups that information, requirements, and plans are not being shared. However, separate management chains would leave the possibility of unresolved conflict.

Finally, the EA could remove any division and place all ABIS/BEC development, operation, and maintenance under one organization. This approach seems to have the greatest potential to alleviate conflict and improve efficiency and effectiveness, and there is also a potential cost saving if functions could be consolidated. Such a change could be initially disruptive, with an organizational concept outside the normal paradigm, but the potential benefits appear to be the greatest of the three options. Statutory requirements to have acquisition functions under the Assistant Secretary of the Army for Acquisition, Logistics and Technology (ASA[ALT]) would have to be addressed, but could probably be done with "dual-hatted" designations and separate reporting chains for acquisition and operational purposes. Improving ABIS/BEC's functioning and ensuring smooth upgrades are critical given COCOM, national-level IC collection and identification, and international cooperation may soon become the largest share of work on biometrics for red and gray forces.[4]

Near-Term Recommendations to Improve Biometrics Activities

Our highest-priority recommendations are intended to address process, priorities, standards, and metrics in the near term.

First, we recommend DoD issue a clear and understandable articulation of the mission for the biometrics enterprise. This should cover biometrics for red, gray, and blue forces; contain an explicit statement about biometrics being an enabling and support tool for warfighter missions; emphasize the need for a functioning end-to-end biometrics capability; and require compliance with biometric data standards. Stating goals for the use of biometrics will help bound most of the future discussions and, in many ways, eliminate longstanding conflicts within DoD. Carrying out this recommendation should include replacing the current Biometrics Enterprise Strategic Plan.

Second, we recommend DoD revise DoDD 8521.01E and issue one or more DoDIs to clarify the objective of the biometrics EA with respect to a complete and functioning end-to-end biometrics cycle. Within this framework, we recommend the updated DoDD ensure that roles and responsibilities have been delineated to confirm responsibility for the cycle and its components and to clarify mediation processes for organizational conflicts. While the DoDD solidified the value of coordinating a biometrics capability within DoD, it preceded a jointly recognized characterization of the objective of the overall biometrics program and the EA. As a result, DoDD 8521.01E relies heavily on organizational designations and technology promotion without ensuring biometrics cycle effectiveness or defining what efficiency means in this context.

Based on our analysis, the updated DoDD should also address the shift from EA "pushing" technology to "pulling" requirements. In particular, we note that conflicts arose between executing the biometrics cycle and the strategic planning required for cycle effectiveness, par-

[4] BIMA Comments, April 2012. BIMA asserts, "This was tried in the past and was unsuccessful. PM's tend to focus on cost and schedule whereas operators focus on performance. When they were combined the developers built their own requirements, approved them, built the system, tested the system with little to [no] input from the operator. The results are a system that continue to fail to meet the operators [sic] and our customers [sic] need." However, we believe a formal PoR would alleviate the issues of a developer working without operator input because requirements from users are necessary for a PoR, along with the rigor and accountability of cost, schedule, and performance parameters.

ticularly when fielding new technology to meet requirements from multiple services.[5] While alternate modifications may improve the effectiveness and the efficiency of the biometrics cycle, we suggest DoD consider the following provisions for inclusion in a revised directive:

1. Require the EA to determine, document, and disseminate root causes for biometrics cycle breakdowns, in coordination with other DoD Components, and require EA to submit proposed and any executed corrective actions or other appropriate actions to the PSA.[6]
2. Designate the PSA as responsible for submitting planning guidance to the EA that will ensure biometrics cycle sufficiency.
3. Require the EA to empower the Executive Manager to coordinate all other EA components. This could entail designating the Executive Manager at an Army organizational level sufficiently high enough to direct and ensure proper execution of biometrics functions.[7]

We noted that current assignments and arrangements exposed conflicting views between the EA and the COCOMs regarding what action is required to ensure an effective biometrics cycle and how requirements should be subsumed in the name of activity efficiency.[8] Therefore, we recommend as follows:

---

[5] DoD, 2008a. Related assignments of the EA in the current directive include:

> E4.1. Provide biometrics research, technology, and information management support to the Chairman of the Joint Chiefs of Staff, the Combatant Commanders, the Military Services, the Defense Agencies, and other DoD Field Activities, as required.
>
> E4.4. Develop and maintain, in coordination with the Heads of the DoD Components as appropriate, the policy, processes, and procedures for the collection, processing, transmission, archiving, caching, and tagging of biometric data and associated intelligence.
>
> E4.5. Develop a DoD biometrics vision and strategy, in concert with the Military Services, Joint Staff, and Office of the Secretary of Defense, for annual submission to the Biometrics PSA via the DoD Biometrics EXCOM.
>
> E4.6. Coordinate DoD Science and Technology (S&T) plans and develop and maintain a DoD biometrics S&T 'roadmap' that clearly demonstrates S&T transition points to acquisition efforts.

[6] The EA can only correct what is within his control, and root causes may reside with other DoD components.

[7] This should not be interpreted as recommending a change in basic Army organization and division of responsibilities.

[8] DoD, 2008a. Areas of possible contention in the current directive include:

> 5.12.8. Coordinate all component biometric requirements with DoD component members of the DoD Biometrics EXCOM.
>
> 5.12.6. Program for and budget sufficient resources to support common enterprise requirements documentation, architecture development, materiel development, test and evaluation, lifecycle management, prototyping, exercises, records management, demonstrations, and evaluations to include efforts at maturing viable technologies and standards.
>
> E4.4. Develop and maintain, in coordination with the Heads of the DoD Components as appropriate, the policy, processes, and procedures for the collection, processing, transmission, archiving, caching, and tagging of biometric data and associated intelligence.

1. Designate the PSA as responsible for adjudicating disagreements between military service priorities and biometrics cycle sufficiency priorities.
2. The language of the current DoDD is vague in parts that could be used as justification for inaction.[9] Therefore, we suggest:
3. The revised directive should strictly avoid language that lists responsibilities without end goals—such as "coordinate all" or "collaborate with"—unless it also provides clear, predefined governance procedures to adjudicate or resolve disagreements.
4. The current DoDD does not adequately address roles and responsibilities regarding biometrics activities that support intelligence. Therefore, we propose:
5. Changes in the directive should explicitly define boundaries of the PSA and EA responsibilities, as well as roles and responsibilities for developing BEI tools and training.

Also, DoD documentation reveals an EA is well suited to execute guidance but is likely not well positioned to represent departmental interests accurately and effectively in external negotiations beyond standards development.[10] We advise the revised directive should re-emphasize the well-understood roles of OSD, namely that the PSA and OUSD (Policy) are exclusively responsible for interagency coordination and international negotiations, unless those responsibilities are explicitly delegated.

Third, we recommend DoD consider a standard architecture and scheme that can be used to transmit biometrics data and match responses. There were understandable wartime pressures to create systems that met needs in the field, but rationale for parallel approaches suggests a failure of responsiveness to user requests. Thus, DoD needs to consider the effi-

---

E4.10. Provide for the establishment and operation of a governance structure, composed of members from the DoD Components and subordinate to the DoD Biometrics EXCOM, to enable the development and execution of common requirements, standards, architectures, and research and development initiatives to support common and joint requirements.

E3.3.1. Provide oversight, preclude duplication of effort, and undertake resolution of issues across DoD biometrics programs. Advise and make recommendations to USD(AT&L) concerning individual biometrics acquisition programs to ensure no unnecessary duplication of effort related to biometrics occurs across the Department.

9 DoD, 2008a. Vague language in the current directive includes:

5.12.8. Coordinate all component biometric requirements with DoD Component members of the DoD Biometrics EXCOM.

E4.3.4. Collaborate with interagency community and other mission partners to facilitate consistent implementation and use of biometric standards.

10 DoD, 2008a. Related assignments of the EA in the current directive include:

E4.3.1. Provide for participation on national and international standards bodies to influence and accelerate standards development.

E4.3.4. Collaborate with interagency community and other mission partners to facilitate consistent implementation and use of biometric standards.

E4.3.5. Provide for the development of tools to facilitate interoperability, both internal and external to the Department of Defense.

ciency and effectiveness of maintaining multiple biometric system architectures. Specifically, DoD should look to transition from the BAT system to web-based portals like those used by SOCOM and recently piloted by the Marines.

We also recommend that DoD prioritize the classified-unclassified cross-domain solutions to eliminate architecture inefficiencies and improve sharing inside and outside DoD. Without clear prioritization, this problem will likely remain unsolved to a degree that satisfies field users, despite the fact that the solution should not be technically insurmountable.

We also recommend that this architecture be designed to ensure match/no match responses are sent to the originator of the request as well as other users of the information. As noted above, this kind of feedback is not a consistent part of the current scheme.

Fourth, we recommend that the biometrics community should inform the discussion of tactical and theater communications but should not be in the business of developing and deploying them. We recognize that the effectiveness of the biometrics cycle depends on timely interactions between the field user and the database. Thus, the DoD biometrics community needs to have a transparent and well-understood process to coordinate and adjudicate requirements with the tactical communications community. Since this process is outside the bounds of the EA, we also recommend that new biometric technology be fully interoperable with current and planned communication and information systems, so that data may be sent by efficient and practical ways. Formal PoRs may address this in part by requiring technical developments to adhere to DoD-wide standards.

Finally, we recommend that the PSA and EA develop metrics to assess both the outcomes and outputs of the biometrics enterprise. Metrics are often cited in DoD, and the biometrics enterprise clearly needs to improve its use of them. EA components have some organizational "metrics" and are collecting some output data, but these do not demonstrate the value of biometrics activities (outcomes) or provide a useful measure of how effectively and efficiently the enterprise is meeting user needs (outputs). DoD should consider a cost-benefit approach to outcome metrics that could facilitate thoughtful consideration of how to spend each marginal dollar effectively, including comparing current outlays to current positive results. For output metrics, DoD should consider more user-focused measures of speed, reliability, and quality of biometric data as compared to some benchmarks. The PSA and EA should work together to use and improve output metrics already being used by the EA.

As noted throughout this report, the DoD biometrics enterprise is at a point of significant change. Large-scale OCOs that had been the focus of support are ending or have already ended, the OCO funding is declining, and base budgets in DoD are being cut. At the same time, intelligence, law enforcement, and international partner desires for biometric capabilities are likely to increase. DoD must now establish goals for the biometrics enterprise, justify the scope and scale of biometrics-related activities to senior leadership (with appropriate measures), and ensure the enterprise meets the user needs effectively and efficiently.

APPENDIX A

Biometrics Enterprise Strategic Plan Goals

- Provide enduring biometrics capabilities to the Department of Defense and its mission partners to enable identification of individuals in support of national security interests.

- **1.0 MILITARY OPERATIONS: Meet warfighting biometric needs of Combatant Commands, Services, and Combat Support Agencies with sufficient timeliness and certitude.**
- 1.1 Employ biometrics collection capabilities with full cooperation, limited cooperation, and no cooperation.
- 1.2 Operate and maintain DoD authoritative sources of biometric information that enable secure and robust interactions between DoD biometrics collection and matching systems and other identity systems.
- 1.3 Share biometrics data among the DoD Components, and with interagency and multinational partners in accordance with law and policy.
- 1.4 Employ biometrics to enable targeting, tracking, and prosecution of known or potential adversaries.
- 1.5 Employ biometrics to enable protection of friendly forces, facilities, and equipment against adversaries and unknown persons.

- **2.0 BUSINESS FUNCTIONS: Integrate biometrics into DoD business functions as required to increase efficiency, effectiveness, and accuracy.**
- 2.1 Clearly identify the areas in business processes where biometrics can provide additional benefit, and determine and prioritize requirements.
- 2.2 Define the strategic, operational, and technical frameworks for biometrics-enabled identity transaction activities. Prioritize modalities and business functions.
- 2.3 Establish and manage DoD Biometrics Enterprise performance metrics.

- **3.0 INSTITUTIONALIZATION: Ensure biometrics are incorporated into standard DoD processes in order to enhance military operations and business functions.**
- 3.1 Develop and implement an approved Enterprise Requirements Strategy that identifies and integrates joint, Service, Combatant Command, and Defense Agency biometrics requirements.

- 3.2 Provide [Planning, Programming, Budgeting and Execution] PPBE guidance for insertion of common biometrics capabilities.
- 3.3 Acquire biometrics capabilities in accordance with DoD 5000 series.
- 3.4 Develop and facilitate adoption of biometrics standards to enable interoperability.
- 3.5 Establish and maintain the DoD Biometrics Enterprise Architecture.
- 3.6 Establish and coordinate a comprehensive S&T program to guide development and transition of technologies that enhance biometrics-enabled capabilities.
- 3.7 Assess, develop, and coordinate DoD policies to enhance employment of biometrics.

- **4.0 UNITY OF EFFORT: Coordinate efforts to achieve unity of purpose for the effective governance and employment of biometrics-enabled capabilities and proactive engagement with joint, interagency, and multinational partners and civilian populations.**
- 4.1 Building upon DoDD 8521.01E, develop effective processes to engage leadership and coordinate efforts across the biometrics community.
- 4.2 Establish the methods and means to execute the roles and responsibilities described within DoDD 8521.01E.
- 4.3 Ensure a continuous strategic planning process that incorporates evolving technologies and innovatively uses capabilities across a range of mission areas.
- 4.4 Engage with interagency and multinational partners to biometrically identify individuals of interest.
- 4.5 Develop and implement strategic communications to inform leaders, stakeholders, and the public of the value of DoD biometrics capabilities.
- 4.6 Develop and implement knowledge management for DoD biometrics.

SOURCE: DoD, *Biometrics Enterprise Strategic Plan: 2008–2015*, August 27, 2008b.

APPENDIX B

Interview Protocol

RAND National Defense Research Institute FY12 "DoD Biometrics Study"

The "DoD Biometrics" study is sponsored by John Boyd (Director, Defense Biometrics). Our aim is to inform the DoD Biometrics program by gathering perspectives of affected stakeholders.

FOR ALL:

[As applicable: I received your name from XX as a stakeholder in the DoD Biometrics program.] To orient the conversation, please tell us about your current position and your personal work responsibilities regarding involving biometrics.

FOR USERS:

Current use

- How is your organization using biometrics currently? We are interested in the operational aspects.

Current organizationally dictated needs

- What are the biometric requirements of your organization? We are looking for official requirements rather than future needs.
- What guidance documents does your organization follow to structure its biometrics program?

Current processes

- How does your organization request new biometrics equipment?
  (Prompts)
  - What types of equipment has your organization requested in the past? Examples?
  - Whom did you request equipment from?
  - How quickly did this organization respond to your request? How quickly did you receive the equipment?
  - In your opinion, overall, did the process for technology acquisition work/not work? Why?

Operational needs

- How well does your organization's current biometrics program meet your operational needs?

- How do you assess this gap?
- In closing, what would you like DoD leadership to know about the current biometrics program?
- What changes, if any, would you recommend?

FOR EA COMPONENTS:

Current activities

- What is your organization's responsibility with respect to biometrics?

Origin of requirements

- What is the source of requirements for your organization's activities and responsibilities?
- How do you obtain them?
- How are they vetted?

Interactions with users

- How do you interact with users of your biometric equipment/information/output?
- How does your organization respond to new requests from the field?
  (Prompts)
  - What types of equipment/support/tools were requested? Examples?
  - Whom did the request come from?
  - How quickly did your organization respond?
  - In your opinion, overall, did the process work/not work? Why?

Operational needs

- How well do your organization's current biometrics activities meet operational needs?
- How do you assess your performance?
- How do you assess any gaps?
- In closing, what would you like DoD leadership to know about the current biometrics program?
  (Prompts)
  - What changes, if any, would you recommend?

APPENDIX C

Metrics

Metrics for Biometrics Activities

Output metrics and outcome metrics serve different purposes. In the world of biometrics, an output metric helps observers know whether the biometrics cycle is running at peak efficiency; that is, whether it processes biometrics at sufficient speed, throughput, and quality to satisfy its customers. Outcome metrics inform the decision of whether to establish and maintain a biometrics system at all by showing the value provided; for instance, given the expenditures and the running cost, is the military finding enough bad guys whose status as bad guys is not otherwise obvious (e.g., they operate under an alias)? If so, one may conclude that the effort expended on biometrics is worthwhile—even if the biometrics cycle is working at less than full efficiency. Overseers may therefore conclude that biometrics efforts are worth investing in, but that perhaps the biometrics cycle needs better management to achieve greater efficiencies, and—everything else being equal—achieve better outcomes.

Conversely, the biometrics cycle may be operating at peak efficiency and just not proving its worth because too few bad guys are being identified as such. The fault, in such a case, may not lie with the biometrics system, which is doing as well as it can, but with the fact that the requirement for biometrics is mismatched with the type of operations carried out by the U.S. military (e.g., its adversaries all wear uniforms that label them correctly). In that (hypothetical) case, one may conclude that having a biometrics system is just not worthwhile, despite the hard work of those who have built and are running it.

The metrics that matter, therefore, depend on how they need to be used. If the aim is to justify the biometrics program, then outcome measures are best. If the aim is to examine how well the biometrics process is being run, then output measures are best. DoD appears to need some biometrics capabilities, even in an era of tight budgets and no large OCOs. But the scale of the biometrics enterprise has not been justified, the current system appears to be capable of operating more efficiently (particularly in the views of warfighter end-users), the EA's management structure has some deficiencies, and nothing suggests matters will be radically different in the near future. Thus, both outcome and output metrics are needed. The PSA should use outcome metrics to report to senior management regarding the value of biometrics and output metrics to monitor and assess the performance of the EA in carrying out its responsibilities.

Of the two types of metrics, outcome metrics are more important because they can demonstrate the overall value, but they are also more difficult to develop. When people begin discussing outcome and value-added metrics, they often give first consideration to return on investment (ROI). This can be a useful frame of reference for a for-profit entity where an investment today can be linked to increases or decreases in revenue tomorrow. However, measuring ROI from DoD biometrics activities—and often DoD work writ large—is more difficult and not necessarily appropriate. For example, DoD and the IC spent a large amount of money trying to locate Osama bin Laden, and they eventually found him. The size of the investment can be measured to some extent, but it is nearly impossible to quantify the value of locating that one person. Moreover, ROI is a useful measure when there is a large initial outlay followed by a payback period extending over subsequent years. DoD invested nearly $3 billion in biometrics from FY07 to FY12 hoping for a near immediate effect in terms of better tracking of detainees, managing access to facilities and services, looking for watch-listed individuals, and more effectively identifying the owners of latent fingerprints.

Instead, DoD should consider a cost-benefit approach to outcome metrics for biometrics, which compares current outlays to current positive results. Specifically, the PSA could consider how many bad actors are identified per dollar spent and compare that to other investments made to identify bad actors.[1] This approach is probably useful because it would be based on some sort of benchmark. A refinement of this would be figuring out the number of bad actors who would not have been identified without biometrics. This would be harder to do but could demonstrate the unique value of biometrics investments.

Even if the value of catching bad actors cannot be quantified, the cost-effectiveness of biometrics expenditures can be gauged by asking: Did biometrics investments provide greater value than other investments that were not necessarily intended to identify bad actors but met similar broader goals? For example, biometrics data have been used to identify IED builders, with the ultimate goal of saving American lives. Could the money invested in biometrics have saved more lives by purchasing more Mine Resistant Ambush Protected vehicles to protect against IEDs, or by operating more unmanned aerial vehicles to reduce the number of ambush attacks against ground forces?

In judging the relative efficacy of biometrics spending, it helps to note that the initial investments have been made, and the resources needed to maintain the existing capability are relatively small by DoD standards. For example, current operating expenses for ABIS (not including system upgrades paid for by PM-Biometrics) and the biometrics examiners at BIMA-West Virginia are approximately $15 million per year, which includes supporting the large deployment of U.S. forces in Afghanistan. Continuing to operate, maintain, and upgrade ABIS at its current level of effort after 2014 would presumably be even less expensive since troop levels will be reduced. However, DoD's future plans include two-way transactions with interagency partner databases, DHS's IDENT and DOJ's IAFIS. Accordingly, outcome measures could evolve to count people of interest who are only identified because their biometric

---

1 BIMA argues that this cannot be done because it does not know the outcome of most biometrics matches.

records are in more than one database. For example, the PSA may already be aware of ABIS data added to the BEWL that resulted in matches submitted from IDENT. Outcome measures, in general, might also be weighted by who has been identified and what apprehending that person contributes to broader national goals, such as reducing the threat from a terrorist group.

Output metrics, on the other hand, are the basis upon which OSD could monitor the efficiency of the biometrics process, and are the means by which OSD could hold the EA to account for its performance. They are particularly valuable if the biometrics enterprise is reconceptualized to have stronger oversight in OSD, while functional responsibility over the biometrics cycle is retained within the EA.

At best, the biometrics cycle should run like a smooth machine, moving biometrics quickly and completely through collection, transmission, matching, and analysis, and then back to warfighters in time to act. In practice, the biometrics cycle (like all human processes) falls short of this ideal. The purpose of metrics is to measure how far short the cycle falls of the ideal. But metrics without some way of measurement are empty, and measurement, in turn, is not free. Thus, the questions of what and how to measure require careful answers.

In theory, three metrics are dominant: speed, reliability, and quality. Speed is how quickly information traverses the system. Reliability is how often information completes the cycle (100 percent being the goal). Quality is the rate at which the matching process produces false positives (a match that identifies two different people as the same person) and false negatives (a person cannot be matched who should be matched). Note that the volume of matches, in and of itself, is not a metric for the biometrics cycle. If those on the BEWL do not present themselves to U.S. forces fast enough, then the number of matches will be correspondingly lower. We shall explore these three metrics in greater detail.

Speed is the cycle time required between when an individual is brought before a biometrics device and when the examiner finds out whether the individual presents a match to someone already in the database. Ironically, this measure cannot stand on its own without some regard for the size and heterogeneity of the database. Is a warfighter better off getting a result from ABIS within five minutes, or getting a result from a combination of ABIS and the DHS's IDENT databases in ten? A great deal depends on the likelihood that such an individual might appear in the latter and not in the former. If an Afghan villager is picked up in an Army sweep, that extra five minutes to check IDENT are probably wasted, but if an urban dweller is picked up by SOCOM in Somalia after having been tipped off by intelligence, those five minutes may well make a big difference.

Further, speed combines several factors: the facility with which the device can be deployed, how fast it works, the transmission speed between the device and the relevant database (or databases, as explained above), the queuing time before the data can be searched against a database automatically, the queuing time before the data can be analyzed by a human examiner if needed, any quality-control checks outside normal processing, and, finally, the time required to get the determination back to the field. BIMA-WV claims it can run at least its part of the cycle within five minutes, but that is, after all, only a part of the cycle. SOCOM decided to go for a direct satellite link to ABIS to demonstrate that they believe enough in the importance of their own requests to invest their own money. They also wanted to get out from under the detours and delays associated with using the BAT-NGIC-BIMA-ABIS data transmission scheme.

As a practical measure, however, a speed metric may not be able to accommodate the time it takes to prepare a biometrics device for use, e.g., unpacking and powering up the device in a field environment. It would also be easier to stop the clock, so to speak, when the confirmation message leaves ABIS, rather than when it arrives at the machine of the requester. Time-stamping the outgoing message from ABIS (correlated to the unique identifier of the initial biometrics file) provides data for determining the biometrics cycle time. The data support a metric that would apply over all biometrics cycles, or to a cycle that is currently segregated to differentiate more urgent requests (perhaps as characterized by priority of submission).

Reliability, in effect, is an extension of the time-metric in the sense that a cycle that never completes is tantamount to a cycle that takes an infinite time to complete. Reliability can be defined in terms of two breakpoints: The data never get to ABIS, or the response from ABIS never gets back to the intended recipient.[2] Unsatisfied requesters can resubmit biometrics under controlled, non-time sensitive conditions such as a detention center, but if the collector and the response recipient are two different individuals, neither may know that the cycle never completed.

Metrics for reliability are easy to state (e.g., percentage), but not necessarily easy to collect. The warfighter may claim that he submitted biometrics, but unless there are the digital equivalents of receipts between the warfighter's device and ABIS (or some junction point on the way to ABIS), the biometrics enterprise can claim to have never received it. As a practical measure, some combination of receipts, for verification, followed by measurements similar to the speed metric should suffice to register completion rates. However, these efforts will not capture incidents when the cycle fails to complete because the biometrics device fails to capture data or fails to transmit them. BIMA is already collecting simpler reliability metrics, such as ABIS downtime and mean time between incidents of ABIS being unavailable, both scheduled and unscheduled.

Quality is a multidimensional metric. It is important to avoid false positives: i.e., two biometrics associated with one individual when they really come from different individuals. It is also important to avoid false negatives: i.e, two biometrics from the same individual that fail to match. But false negatives may also result when the database of biometrics is too small. (More typically, false positives and negatives occur during attempts to match a fully enrolled biometric record to latent biometrics, because the latter is almost always a fraction of a full biometric. Conversely, latent matching has the advantage of having people in the loop.) Finally, if the biometrics samples themselves are indistinct (e.g., smudged), incomplete (e.g., the flat side

---

[2] As noted elsewhere, match/no match responses from ABIS already do not make it back to the requester except when using the BAT architecture.

of the finger rather than the whole finger), or nonstandard, then it may not match against a current or future database entry.

A challenge to measuring quality is the lack of a standard of comparison, notably the false negative and the false positive rates. How can one know, for instance, that two people have been confused with one another based on extant evidence, until later evidence suggests that two individuals were involved? If the evidence does come in later, one might conclude that prior biometrics efforts had quality problems, but it will not be possible to make such judgments about current biometrics efforts until later confirmation or refutation becomes available. One metric may be to keep track of matches and (1) which of them are later rejected based on subsequent evidence, or (2) which of them could have reasonably been made at an earlier opportunity but were not. Such metrics may be indicative, but not necessarily definitive.

Conversely, it is easier to judge whether a biometric has simply not been captured very well because there are standards that can be applied to quality per se that do not require the later application of actual use to evaluate. An adequate metric may be constructed by sampling the universe of biometrics in ABIS and testing it against the de minimum standards of other organizations in the biometrics business. Alternatively, if BIMA decides to reject biometrics of insufficient quality in the future, rejection rates may be a quality test of at least part of the biometrics cycle. The FBI, for example, can do this because it usually gets biometrics from people who have been arrested and, if not released shortly thereafter, are in custody; local law enforcement can also make multiple collections to meet quality standards.[3]

Complicating these output metrics is the fact that not every biometric is equally important. A special operations team carrying out a mission against possible terrorists may have a keen interest in determining quickly, completely, and accurately whether it caught one while on the mission. Conversely, someone collecting biometrics on an individual in extended detention may be more flexible, and could, if necessary, submit information into the system multiple times (although doing so is undesirable). Similarly, the odds of a no-match turning into a match can vary widely depending on what population pool the specific biometric came from, e.g., Afghan detainees compared with foreign nationals entering the United States. As a practical matter, a system that can rapidly handle high-priority requests, even if it has to bump low-priority requests, may be more useful to national defense than one that handles all requests equally with a faster average handling time.

Both outcome and output metrics are critical for the PSA going forward to justify continued biometrics investments and to monitor the EA's performance, respectively. The EA collects data for output metrics, such as speed, completeness, and conformance to standards. Developing adequate outcome metrics will require additional effort.

---

[3] The FBI also collects fingerprints for applicants (e.g., for federal jobs) who are likely to return if the first set is deemed inadequate.

Metrics Used by BIMA

An examination of the metrics actually used by BIMA suggests a heavy concentration of input/process metrics at the expense of output or outcome metrics. Even these appear to have been developed relatively recently despite multiple years of discussion.

BIMA's May 1, 2012, report, *Biometrics Enterprise Strategic Plan (BESP) Metrics* lists 54 separate metrics, divided evenly between category 1 metrics, where all the data required to fill out the metric comes from BIMA, and category 2 metrics, where BIMA would need help from other entities. Of the 27 category-1 metrics, 10 measure BIMA's ability to construct metrics, and only 15 had any data associated with them, suggesting that the other 12 are notional metrics at this point. Similarly, 13 of the 27 category-2 metrics had data associated with them. The data appear to be overrepresented by indications of BIMA effort rather than actual accomplishments. For instance, metric 3.4F ("develop and facilitate adoption of biometrics standards to enabled interoperability") is rated as green, or 100 percent. This declaration is reached by calculating the number of meetings held (in this case two meetings) divided by the number of meetings scheduled (also two meetings).

Of the 54 metrics, only two of them say anything useful about the biometrics enterprise (as opposed to BIMA itself). One is metric 4.1a, which indicates the "total time from [the] ingestion of biometrics at [the] transaction manager to the creation of a match report" batched in terms of the threshold definition of priority (5 minutes for priority 1, and 30, 60, and 240 minutes for priorities 2, 3, and 4, respectively). If the data are credible, they suggest that the processing time is independent of priority level; the lowest average time is for priority 3. The other is more of an outcome metric, the number of watch-list hits per week.[4]

Perhaps most importantly, the metrics are all based on the 2008 BESP, which merits reevaluation at best and is outdated at worst, given the changes in DoD operations since 2008 (and further changes expected after 2014).

---

[4] One week was especially productive, yielding four times more watch-list hits than the next best week and ten times as many as the average of the other 29 weeks surveyed, which raises the question of whether there was more bad actor activity or a lag in data transmission.

APPENDIX D

Reference List by Document Type

Acquisition Documents

- U.S. Central Command Joint Urgent Operational Need Statement (JUONS) for a Joint Biometrics Solution in Support of Operations, 2005
- U.S. Central Command JUONS for Biometrically Based Access Control System, 2006
- U.S. Joint Forces Command, Biometrics, Quick Look-Capabilities Based Assessment, 2006
- Hand-held Interagency Identity Detection Equipment (HIIDE) System Evaluation Report, BFC-TE-RT-06-023, 2007, U.S. Army Biometrics Task Force.
- Body of Biometrics in Support of Identity Management, Joint Capabilities Document, 2008
- Capability Development Document (CDD) for Personnel Identification Version One (PI V1) Revision 1.0, 2008
- DoD Biometrics Acquisition Decision Memorandum, September 2, 2008
- Regina Begliutti, Biometrics Enterprise Capability (including DoD authoritative database) Analysis of Alternatives–Guidance, OSD(PA&E) 2008.
- Joint Requirements Oversight Council, Memorandum: Identity Dominance System Capability Development Document, JROCM 181-08, 2008.
- Thomas Dee, Defense Biometrics: Material Development Decision Brief: Integrated Biometrics Programs, Biometrics OSD AT&L (DDR&E), 2009.
- Initial Capabilities Document for Biometric Enabled Intelligence (BEI) version 0.89, 2009.
- Biometrics Tactical Collection Device (TCD) and Biometrics Enterprise Capability (BEC) Analysis of Alternatives, 2010.
- CENTCOM Joint Urgent Operation Need Statement for a Joint Biometrics Solution in Support of Operations.
- Approval for Biometrics in Support of Identity Management ICD. JROCM 210-08
- Approval for Navy IDS CDD. JROCM 181-08

BIMA Annual Reports

- Biometrics Task Force Annual Report, Biometrics Task Force, DoD, 2007.
- Biometrics Task Force Annual Report, Biometrics Task Force, DoD, 2008.
- Biometrics Task Force Annual Report, Biometrics Task Force, DoD, 2009.

- Biometrics Identity Management Agency Annual Report, Biometrics Identity Management Agency, DoD, 2010.
- Biometrics Identity Management Agency Annual Report, Biometrics Identity Management Agency, DoD, 2011.
- Biometrics: The Army As Executive Agent for the DoD, U.S. Army, 2008.

DoD Biometrics Program Information

- John Boyd, Naval Identity Management Coordination Meeting, U.S. Navy, DoD, 2009.
- Thomas Dee, The State of DoD Biometrics. 2010, OSD(AT&L), DDR&E
- The Road from Quick Reaction Capability to a Program of Record. 2011,PM DoD Biometrics and TRADOC Capability Manager–Biometrics and Forensics
- Dalton Jones, Future of Defense Biometrics, Forensics, and Identity Intelligence, OUSD(I), DUSD (I&S), 2011.
- John Boyd, OSD (AT&L), Assistant Secretary of Defense for Research and Engineering, Looking Back and Moving Forward … DoD Biometrics, 2011.
- Biometric EXCOM, 2011.

DoD Biometrics Related Legislation

- H.R. 6523, 111th Congress: Ike Skelton National Defense Authorization Act for Fiscal Year 2011, 2010.

DoD Biometrics Related Policy

- Rudy de Leon, Deputy Secretary of Defense, Executive Agent for the DoD Biometrics Project, 2000.
- John Stenbit, Assistant Secretary of Defense,DoD Strategic Plan for Biometrics, 2002.
- DoD Directive 5101.1, DoD Executive Agent, May 2003.
- John Stenbit, Assistant Secretary of Defense, Designation of Biometrics as ACAT 1AM, 2003.
- Paul Wolfowitz, Deputy Secretary of Defense, (DoD) Biometrics Enterprise Vision, 2003.
- John Stenbit, Chief Information Office, DoD Identity Management Senior Coordinating Group (IMSCG). 2004.
- John Stenbit, DoD Chief Information Office, DoD Compliance with the Internationally Accepted Standard for Electronic Transmission and Storage of Fingerprint Data from "Red Force" Personnel. 2004.
- John Stenbit, Assistant Secretary of Defense, Development of an Identity Management (IdM) Vision for the Biometrics Program. 2004.
- Deputy Secretary of Defense, DoD Policy for Biometric Information for Access to U.S. Installations and Facilities in Iraq, OSD 12922-5, 2005.
- Interim Executive Agent Guidance on Ownership of Biometric Data Submitted to and Stored in the DOD Automated Biometric Identification System (ABIS), 2005.
- Gordon England, Deputy Secretary of Defense, Defense Biometrics. 2006.

- DoD Directive 8521.01E, DoD Biometrics. 2008.
- Designation of the Executive Manager for DoD Biometrics, 2008.
- Capstone Concept of Operations for DoD Forensics, 2008.
- DoD, DoD Biometrics Enterprise Strategic Plan: 2008-2015, 2008.
- DoD, DoD Identity Management Strategic Plan, 2009.
- David Wennergren, DoD Deputy Chief Information Officer, (DoD) Identity Management Strategic Plan, 2009.
- The Office of the Assistant Secretary of Defense for Networks and Information Integration/DoD Chief Information Officer, DoD Privilege Management Roadmap, 2010.
- John M. McHugh, Headquarters, Department of the Army, GO 2010–06, Redesignation of the Biometrics Task Force as the Biometrics Identity Management Agency, 2010.
- CENTCOM Biometric Identification System for Access (BISA) CONOPS.
- EUCOM Concept of Operations for the Use of Biometric Technologies.

Executive Government Biometric Related Policy

- George Bush, Integration and Use of Screening Information, Homeland Security Presidential Direct/HSPD-6, HSPD-6, 2003.
- George Bush, Comprehensive Terrorist-Related Screening Procedures, The White House, HSPD-11, 2004.
- George Bush, Policy for a Common Identification Standard for Federal Employees and Contractors, The White House, HSPD-12, 2004.
- George Bush, Biometrics for Identification and Screening to Enhance National Security, The White House, HSPD-24, 2008.
- National Science and Technology Council: Subcommittee on Biometrics and Identity Management National Science and Technology Council Policy for Enabling the Development, Adoption and Use of Biometric Standards, 2008.

Executive Government Biometric Related Reports

- National Science and Technology Council: Subcommittee on Biometrics, The National Biometrics Challenge, 2006.
- National Science and Technology Council, Biometrics in Government Post-9/11: Advancing Science, Enhancing Operations, 2008.
- National Science and Technology Council Subcommittee on Biometrics and Identity Management, Executive Office of the President of the United States, Identity Management Task Force Report, 2008.
- The President's National Security Telecommunications Advisory Committee, NSTAC Report to the President on Identity Management Strategy, 2009.
- National Research Council, Whither Biometrics Committee, Biometric Recognition: Challenges and Opportunities, 2010.
- National Science and Technology Council: Subcommittee on Biometrics and Identity Management, The National Biometrics Challenge, 2011.

Reports on DoD Biometrics

- Office of the Under Secretary of Defense for Acquisition Technology and Logistics, Defense Biometrics, 2007.
- Jeffrey L. Caddell, Privacy & Security Law Report, DoD Biometrics in the Age of Terror, 2008.
- Defense Management: DoD Needs to Establish Clear Goals and Objectives, Guidance and a Designated Budget to Manage its Biometrics Activities, GAO-08-1065, 2008.
- Defense Biometrics: DOD Can Establish More Guidance for Biometrics Collection and Explore Broader Data Sharing, GAO-09-49, 2008.
- Terry C. Heuring, Serena Chan, Dennis F. DeRiggi, Keith L. Green, Lance G. Joneckis, William M. Knarr, Kwan S. Kwok, Tara A. McGovern, Lawrence B. Morton, Robert A. Stimets, The Department of Defense Biometric Enterprise, Institute for Defense Analysis, IDA Paper P-4416 (DRAFT Final), 2008.
- Office of the Under Secretary of Defense for Acquisition, Technology, and Logistics, Operations Research Applications for Intelligence Surveillance and Reconnaissance (ISR), 2009.
- Defense Biometrics: DoD Can Better Conform to Standards and Share Biometric Information with Federal Agencies, GAO 2011.
- Charles R. Spaulding, W. Scott Gibson, Stephen F. Schreurs, Duane L. Linsenbardt, Antonio DeSimone, Systems Engineering for Complex Information Systems in a Federated, Rapid Development Environment, Johns Hopkins APL Technical Digest, 2011.
- Defense Biometrics: Additional Training for Leaders and More Timely Transmission of Data Could Enhance the Use of Biometrics in Afghanistan, GAO-12-442, 2012.
- Team ANSER, DoD Biometrics Capabilities Integration and Architecture Development Initiative: "As-Is" Business Process, CDRL A008, Biometrics Task Force, September 29, 2009.
- Team ANSER, DoD Biometrics Capabilities Integration and Architecture Development Initiative: "To-Be" Business Process Model (CDRL A009), Biometrics Task Force, August 31, 2009.
- Team ANSER, DoD Biometrics Capabilities Integration and Architecture Development Initiative: 2010, 2015, 2025 Business Process Prototype Model Documentation (CDRL A010), Biometrics Task Force, August 31, 2009.

Bibliography

Begliutti, Regina, *Biometrics Enterprise Capability (including DoD authoritative database) Analysis of Alternatives*, Office of the Secretary of Defense, Program Analysis and Evaluation (OSD[PA&E]), February 26, 2008.

Biometrics and Identity Management Agency website, n.d. As of July 1, 2012: http://www.biometrics.dod.mil/

———, Annual Report, 2010.

———, Annual Report, 2011.

Biometrics Task Force—*See* U.S. Department of Defense Biometrics Task Force.

Boyd, John, Project Manager for Naval Identity Management Capability, "Naval Identity Management Coordination Meeting," with U.S. Navy, Department of Defense, September 10, 2009.

———, Director, Defense Biometrics & Forensics, Office of the Secretary of Defense, Acquisition, Technology and Logistics, *Looking Back and Moving Forward … DoD Biometrics*, September 27, 2011.

Caddell, Jeffrey L., "DoD Biometrics in the Age of Terror," Privacy & Security Law Report, Vol. 07, No. 23, June 9, 2008.

De Leon, Rudy, Deputy Secretary of Defense, "Executive Agent for the Department of Defense (DoD) Biometrics Project," memorandum for Secretaries of the Military Departments; Chairman of the Joint Chiefs of Staff; Under Secretaries of Defense; Director, Defense Research and Engineering; Assistant Secretaries of Defense; General Counsel of the Department of Defense; Inspector General of the Department of Defense; Director, Operational Test and Evaluation; Assistants to the Secretary of Defense; Director, Administration and Management; Directors of the Defense Agencies and Directors of the DOD Field activities, Washington, D.C., December 27, 2000.

Dee, Thomas, *Defense Biometrics: Material Development Decision Brief: Integrated Biometrics Programs*, Biometrics Office of the Secretary of Defense, Under Secretary of Defense for Acquisition, Technology and Logistics (Director of Defense Research and Engineering)(OSD AT&L [DDR&E]), 2009.

———, *The State of DoD Biometrics*, Biometrics Office of the Secretary of Defense, Under Secretary of Defense for Acquisition, Technology and Logistics (Director of Defense Research and Engineering)(OSD AT&L [DDR&E]), 2010.

Defense Acquisition Management Information Retrieval, *Major Automated Information System Annual Report*, Department of Defense Biometrics Project Management Office, January 19, 2012.

Defense Science Board Advisory Group on Defense Intelligence, *Operations Research Applications for Intelligence Surveillance and Reconnaissance (ISR)*, Washington, D.C.: Office of the Under Secretary of Defense for Acquisition, Technology, and Logistics, January 2009.

Director of Defense Research and Engineering, "Capstone Concept of Operations for DoD Forensics," Assistant Secretary of Defense for Research and Engineering, August 5, 2008.

DoD—*See* U.S. Department of Defense.

England, Gordon, Acting Deputy Secretary of Defense, "DoD Policy for Biometric Information for Access to U.S. Installations and Facilities in Iraq," memorandum for Secretaries of the Military Departments, Chairman of the Joint Chiefs of Staff, Under Secretaries of Defense, Commanders of the Combatant Commands, and Directors of the Defense Agencies, Washington, D.C., July 15, 2005.

———, Deputy Secretary of Defense, "Defense Biometrics," memorandum, Washington, DC, October 4, 2006.

GAO—*See* U.S. Government Accountability Office.

Geneva Convention Relative to the Treatment of Prisoners of War, August 12, 1949, 6 U.S.T. 3316, 75 U.N.T.S. 135.

Greenfield, Victoria A., Valerie L. Williams, Elisa Eiseman, *Using Logic Models for Strategic Planning and Evaluation: Application to the National Center for Injury Prevention and Control*, Santa Monica, Calif.: RAND Corporation, TR-370-NCIPC, 2006. As of August 7, 2012:
http://www.rand.org/pubs/technical_reports/TR370.html

Heuring, Terry C., Serena Chan, Dennis F. DeRiggi, Keith L. Green, Lance G. Joneckis, William M. Knarr, Kwan S. Kwok, Tara A. McGovern, Lawrence B. Morton, and Robert A. Stimets, *The Department of Defense Biometric Enterprise*, Alexandria, Va.: Institute for Defense Analysis, IDA Paper P-4416, December 2008.

Homeland Security Presidential Direct/HSPD-6, "Integration and Use of Screening Information," Washington, D.C.: The White House, September 16, 2003.

———/HSPD-11, "Comprehensive Terrorist-Related Screening Procedures," Washington, D.C.: The White House, August 27, 2004a.

———/HSPD-12, "Policy for a Common Identification Standard for Federal Employees and Contractors," Washington, D.C.: The White House, August 27, 2004b.

———/HSPD-24, "Biometrics for Identification and Screening to Enhance National Security," Washington, D.C.: The White House, June 5, 2008.

Ike Skelton National Defense Authorization Act for Fiscal Year 2011, Pub. L. No. 111-383 Stat 4137 (2011).

Joint Requirements Oversight Council (JROCM), "Identity Dominance System Capability Development Document," Washington, D.C.: Joint Staff, September 8, 2008a.

——— JROCM 181–08, "Approval for Navy IDS CDD," Joint Staff, 2008b.

——— JROCM 210–08, "Approval for Biometrics in Support of Identity Management *Initial Capabilities Document*," Joint Staff, 2008c.

Jones, Dalton, *Future of Defense Biometrics, Forensics, and Identity Intelligence*, Office of the Under Secretary of Defense, Deputy Under Secretary of Defense, Installations and Environment (OUSD[I], DUSD [I&S]), 2011.

McHugh, John M., *Redesignation of the Biometrics Task Force as the Biometrics Identity Management Agency*, Headquarters, Department of the Army, GO 2010–06, 2010.

———, Secretary of the Army, "Future of the Biometrics Identity Management Agency," memorandum, Washington, D.C., March 12, 2012.

National Research Council, *Whither Biometrics Committee, Biometric Recognition: Challenges and Opportunities*, Washington, D.C.: National Academies Press, 2010.

National Science and Technology Council Subcommittee on Biometrics and Identity Management, *The National Biometrics Challenge*, August 2006.

———, *NSTC Policy for Enabling the Development, Adoption and Use of Biometric Standards*, September 7, 2007.

———, *Identity Management Task Force Report*, 2008a.

———, *Biometrics in Government Post-9/11: Advancing Science, Enhancing Operations*, Washington, D.C.: August 2008b.

———, *The National Biometrics Challenge*, August 2011.

Office of the Assistant Secretary of Defense for Networks and Information Integration/DoD Chief Information Officer, *Department of Defense Privilege Management Roadmap*, Department of Defense, January 6, 2010.

Office of the Under Secretary of Defense for Acquisition Technology and Logistics, *Defense Biometrics*, Washington, D.C.: Department of Defense, March 2007.

Patton, Margaret, *Biometric Identification System for Access (BISA) Configuration Control Charter and Standard Operating Procedure (SOP) Version 2.1*, DCN: I3-PM-CCB-004 U.S. Army PM-Biometrics, April 2011.

President's National Security Telecommunications Advisory Committee, *NSTAC Report to the President on Identity Management Strategy*, May 21, 2009.

Program Manager for Biometrics and TRADOC Capability Manager–Biometrics and Forensics, *The Road from Quick Reaction Capability to a Program of Record*, 2011.

Spaulding, Charles R., W. Scott Gibson, Stephen F. Schreurs, Duane L. Linsenbardt, and Antonio DeSimone, "Systems Engineering for Complex Information Systems in a Federated, Rapid Development Environment," Johns Hopkins APL Technical Digest, Vol. 29, No. 4, March 2011.

Stenbit, John, Pentagon Chief Information Officer, "DoD Strategic Plan for Biometrics," memorandum to Secretaries of the Military Service; Chairman of the Joint Chiefs of Staff; Under Secretaries of Defense; Director, Defense Research and Engineering; Assistant Secretaries of Defense; General Counsel of the Department of Defense; Inspector General of the Department of Defense; Director, Operational Test and Evaluation; Assistants to the Secretary of Defense; Director, Force Transformation; Director, Net Assessment; Director, Administration and Management; Directors of the Defense Agencies; and Directors of the DoD Field Activities, Washington D.C, June 28, 2002.

———, "Designation of Biometrics as ACAT 1AM," memorandum for Army Acquisition Executive, Washington, D.C., May 15, 2003.

———, "Identity Management Senior Coordinating Group (IMSCG)," memorandum for Chief Information Office Executive Board Members, Wahington D.C., January 12, 2004a.

———, "Department of Defense (DoD) Compliance with the Internationally Accepted Standard for Electronic Transmission and Storage of Fingerprint Data from 'Red Force' Personnel," memorandum for Secretaries of the Military Departments; Chairman of the Joint Chiefs of Staff; Under Secretaries of Defense; General Counsel of the Department of Defense; Director, Operational Test and Evaluation; Director, Defense Research and Engineering; Inspector General of the Department of Defense; Assistants to the Secretary of Defense; Director, Administration and Management; Director, Program Analysis and Evaluation; Director, Net Assessment; Director, Force Transformation; and Directors of the DoD Field Activities, Washington, D.C., February 2, 2004b.

———, "Development of an Identity Management (IdM) Vision for the Biometrics Program," memorandum for Army Acquisition Executive and Army Chief Information Officer, Washington, D.C., February 20, 2004.

Team ANSER, "2010, 2015, 2025 Business Process Prototype Model Documentation (CDRL A010)," Biometrics Integration Analysis Initiative, Biometrics Task Force, August 31, 2009a.

———, " 'To Be' Business Process Model (CDRL A009)" DoD Biometrics Integration Analysis Initiative, Team ANSER, Biometrics Task Force, August 31, 2009b.

———, "'As-Is' Business Process (CDRL A008)," Biometrics Integration Analysis Initiative, Team ANSER, Biometrics Task Force, September 29, 2009c.

U.S. Army, *Biometrics: The Army as Executive Agent for the Department of Defense*, February 26, 2008.

———, *Army Regulation 71-9: Warfighting Capabilities Determination*, Washington, D.C.: U.S. Army, December 2009. As of July 1, 2012:
http://www.apd.army.mil/pdffiles/r71_9.pdf

U.S. Central Command, *Joint Urgent Operational Need Statement (JUONS) for a Joint Biometrics Solution in Support of Operations*, 2005.

———, *Joint Urgent Operational Need Statement (JUONS) for Biometrically Based Access Control System*, 2006.

———, *Biometric Identification System for Access Concept of Operations*, April 15, 2008.

U.S. Department of Commerce, National Institute of Standards and Technology, *NIST Special Publication 500–271: Information Technology: American National Standards for Information Systems—Data Format for the Interchange of Fingerprint, Facial, & Other Biometric Information – Part 1*, ANSI/NIST-ITL 1-2007, Gaithersburg, Md.: Information Access Division, Information Technology Laboratory, National Institute of Standards and Technology, May 2007.

U.S. Department of Defense, "DoD Executive Agent," DoD Directive 5101.1, May 9, 2003.

———, "Department of Defense Biometrics," DoD Directive 8521.01E, February 21, 2008a.

———, *Biometrics Enterprise Strategic Plan: 2008–2015*, August 27, 2008b.

———, DoD Identity Management Strategic Plan, April 2009.

———, *Biometric EXCOM*, October 21, 2011.

———, *Sustaining U.S. Global Leadership: Priorities for 21st Century Defense*, January 2012a.

———, *Capability Development Document (CDD) for Joint Personnel Identification, Version 2*, March 5, 2012b.

U.S. Department of Defense Biometrics Management Office, "Interim Executive Agent Guidance on Ownership of Biometric Data Submitted to and Stored in the DOD Automated Biometric Identification System (ABIS)" memorandum, July 15, 2005.

U.S. Department of Defense Biometrics Task Force, *Annual Report*, 2007.

———, *Annual Report*, 2008.

———, *Annual Report*, 2009.

U.S. European Command Unit 30400 APO AE 09130, *Concept of Operations for the Use of Biometric Technologies*, n.d.

U.S. Government Accountability Office, *Defense Management: DoD Needs to Establish Clear Goals and Objectives, Guidance and a Designated Budget to Manage Its Biometrics Activities*, Washington, D.C., GAO-08-1065, September 2008a.

———, *Defense Biometrics: DoD Can Establish More Guidance for Biometrics Collection and Explore Broader Data Sharing*, Washington, D.C., GAO-09-49, October 2008b.

———, *Defense Biometrics: DoD Can Better Conform to Standards and Share Biometric Information with Federal Agencies*, Washington, D.C., GAO-11-276, March 2011.

———, *Defense Biometrics: Additional Training for Leaders and More Timely Transmission of Data Could Enhance the Use of Biometrics in Afghanistan*, Washington, D.C., GAO-12-442, April 2012.

U.S. Joint Forces Command, *Biometrics, Quick Look—Capabilities Based Assessment*, 2006.

Wennergren, David, DoD Deputy Chief Information Officer, "Department of Defense (DoD) Identity Management Strategic Plan," memorandum for DoD Identity Protection and Management Senior Coordinating Group (IPMSCG), Washington, D.C., August 20, 2009.

Wolfowitz, Paul, Deputy Secretary of Defense, "Department of Defense (DoD) Biometrics Enterprise Vision," memorandum to Secretaries of the Military Departments; Chairman of the Joint Chiefs of Staff; Under Secretaries of Defense; Assistant Secretaries of Defense; General Counsel of the Department of Defense; Director, Operational Test and Evaluation; Inspector General of the Department of Defense; Assistants to the Secretary of Defense; Director, Administration and Management; Director, Program Analysis and Evaluation; Director, Net Assessment; Director, Force Transformation; Directors of the Defense Agencies; Directors of the DoD Field Activities, Washington, D.C., August 25, 2003.